NEPAL EARTHQUAKE

KEVIN BUBRISKI

WITH

Prawin Adhikari • Anil Chitrakar • Muna Gurung • Tsering D Gurung
Niranjan Kunwar • Smita Magar • Pramod Mishra • Pranaya SJB Rana
Avishek Shrestha • Rabi Thapa • Kesang Tseten • Prateebha Tuladhar

FOREWORD BY
Manjushree Thapa

EDITED BY
Deepak Thapa

Himal Books

ISBN: 978 9937 597 69 2

Book design by Norbo Lama.

Himal Books
Himal Kitab Pvt Ltd
521 Narayan Gopal Sadak, Lazimpat
Kathmandu–2, Nepal
Tel: +977-1-4422794
info@himalbooks.com | www.himalbooks.com

Printed in Nepal by
Wordscape The Printer Pvt Ltd
Bhaisepati, Lalitpur, Nepal

Price: Rs 5000/-

Contents

Foreword – Manjushree Thapa ix

Editor's note – Deepak Thapa xiii

Preface: Narratives of a disaster – Kevin Bubriski xv

Acknowledgements xvii

Kathmandu 1

A long time coming – Prawin Adhikari 10

Swayambhunath 15

Nepali resilience – Anil Chitrakar 28

Naya Bazaar 31

How to build a home that lasts – Muna Gurung 44

Gongabu Bus Park 49

Memories of a disaster – Tsering D Gurung 64

Boudha 67

Ruptures and revelations – Niranjan Kunwar 82

Sankhu 87

Together we survive – Smita Magar 94

Bhaktapur 99

Mourning and resilience – Pramod Mishra 116

Patan 121

Unmoored – Pranaya SJB Rana 132

Khokana 135

BudhiGandaki – Avishek Shrestha 142

Bungamati 147

Riding the rubble – Rabi Thapa 160

Chautara 165

Filming Langtang: In the aftermath of the earthquake – Kesang Tseten 172

Charikot 177

The refuge of tremors – Prateebha Tuladhar 184
Dolakha 189
Photographer's journal – Kevin Bubriski 196
The contributors 216

Foreword

— Manjushree Thapa

Taken in by the apparent sturdiness of the world, we lose sight of the fact that life is, in truth, fragile. *Nepal Earthquake* reminds us of this fragility, turning our attention back to a time when the earth shook, leaving a trail of rubble and grief in Nepal.

This book serves as a valuable memory project for a society prone to forgetfulness. In his photographs and in his journal from the aftermath of the April 2015 earthquake, Kevin Bubriski invites us to remember all the lives the earthquake took, and all of the dreams it disrupted: everyday dreams of nourishment, shelter, clothing. Community. Healthcare. Psychological wellbeing. Love. Knowledge: science, history. Philosophy. Wonder and amazement. Art. The dream of a rich and fulfilling life.

Bubriski also invites us to dwell on just how painstakingly Nepal built back from the rubble and grief. In the immediate aftermath of the earthquake, everyone scrambled to save and treat victims, to build temporary shelters from all materials at hand, to distribute rations, and to ensure each others' survival. Once this painful rescue work came to an end, Nepal launched into the arduous process of recovery. Organised in informal groups or formal institutions, people sifted through the wreckage, sometimes protected by hard hats, boots, gloves, and masks, but most times entirely unprotected, using bare hands to clear broken bricks, fallen wooden beams, exposed rebar, snapped electrical wires, shattered roofs, buried statues, and personal belongings, and to heroically salvage a collective humanity.

This is the moment that *Nepal Earthquake* memorialises: a Nepal still reeling from aftershocks and trauma, when rebuilding was still a distant dream.

The 2015 earthquake proved definitional for the nation. The devastation—and the political gains to be made from rebuilding—broke a prolonged stalemate among the governing parties, who set about to hastily promulgate a divisive new constitution. The areas hardest hit by the earthquake—rural Nepal—faced immense bureaucratic hurdles to access aid, while those least affected quickly moved on. The natural disaster revealed the inequalities structured into Nepal. The essays in *Nepal Earthquake* reflect on all that followed, in its full messiness, with hope and heartbreak still playing out, years later.

How do you build a life, a society—and a nation—on volatile grounds? How do you 'build back better', in the absence of political leadership?

Some of the essayists, such as Anil Chitrakar, see cause for great optimism. The social capital that Nepal has earned over the years brings the country goodwill from around the world. And Nepalis have the knowhow and skill to survive, and even thrive, amid adversity. Chitrakar focuses on the unity Nepalis showed after the earthquake, which is part of the beauty of this land.

Others essayists are more conflicted.

'They told us that there was a lesson here somewhere, in the rubble of homes and the detritus of hearts,' Pranaya Rana writes. 'We were supposed to come out of it better, more empathetic, less fatalistic. I am not sure we learnt anything. I'm not sure there was even a lesson there.'

From his essay, and from the others who observed and participated in rebuilding efforts on the ground, we see the heartache involved in communities trying to piece their lives back together. Prawin Adhikari makes civil society's response to the devastation in the countryside vivid and palpable. Kesang Tseten questions his own judgement as he follows the Langtang community for a year, filming in the background as they take charge of rebuilding their wrecked villages and social networks. Niranjan Kunwar, too, turns the questions on do-gooders as he engages with Tamang communities who experience structural violence as a matter of course, and also in their effort to rebuild from the earthquake.

Other essayists write with an even more personal voice. In her moving essay, Muna Gurung investigates the meaning of home, and examines, in intimate, familial terms, how one comes to have a safe, sturdy home. Rabi Thapa offers us a kaleidoscope onto the earthquake's aftermath, with vignettes of people responding, around the world, to the literal and figurative instability of post-earthquake Nepal.

Prateebha Tuladhar captures the numbness with which she and other journalists operated in the immediate aftermath of the earthquake, and the suppressed trauma that lingers on, even years afterwards. And Smita Magar writes of the imperative, and at times overwhelming challenge, of doing good in a community in crisis.

These personal voices are countered by descriptive essays. So much of what the world learnt about the earthquake came from the reporting by Nepali journalists. Tsering D Gurung chronicles what it was like to work in a newsroom in crisis while everything—the world itself—felt like so much ephemera. Pramod Mishra reflects on the deep social fissures that the earthquake revealed, in light of the divisive politics that followed. He identifies the post-earthquake moment as one that gave rise to a new, yet familiar, nationalism in Nepal: a nationalism based on the supremacy of the traditional elite.

No matter which side of the political divide Nepalis fall on, there is, in the end, the land to reckon with: a land of great beauty and great volatility. Nepali society is, and has always been, built on geologically unstable ground. Avishek Shrestha's tender recollection of growing up by the Budhi Gandaki river brings us back to this truth: that Nepalis who pay close attention can almost sense the land in their bones.

All of these essays, and Bubriski's keen observations, come together in *Nepal Earthquake* to testify to the beauty of Nepal. A full analysis of the earthquake—of the manmade disasters that followed the natural one—is yet to take place. Meanwhile, here is a memorial.

Nepal Earthquake is not a book to rush through, but one to reflect over. How long can those struts hold up that temple? The book raises questions. How many weeks did that family live under tarpaulins? Look at that building leaning on another one. Did its inhabitants survive? Look at the workers standing on the rubble in open rubber sandals. What of their former lives was this community able to save? There is so much goodness in evidence here, so much determination and grit. A Buddha statue stands untouched amid a heap of broken bricks. A girl in a uniform is headed to school. Women plant paddy in the fields, planning for another harvest. Life goes on despite all of the grief, all of the loss.

So, let us sit with *Nepal Earthquake*, hearts open, allowing ourselves to think and feel and ask questions. Has Nepal built back better? Can it still do so? What have we done so far? What are we going to do now?

Editor's note

— Deepak Thapa

Till recently, whenever there was talk of earthquakes, the *Nabbe Saalko Bhuichalo* (Earthquake of the Year 90) would inevitably crop up. That cataclysmic event from the year 1990 in the Bikram calendar (1934 CE) had attained mythic status in Nepal since most of what one learnt about it was through the oral accounts of those who had either lived through it or had heard stories about what had happened. Photographic evidence did exist, but for the most part the means to publicise them did not. It was only in the previous couple or so decades, with the growth of the mass media and the flowering of an indigenous publishing industry—not to mention the occasional exhibition—that ordinary folks were able to see for themselves the ravages of that earthquake.

Perhaps because it has not been that long, the 2015 earthquake has not seeped into the national consciousness in the same way the one 80 years earlier had. Perhaps it never will, considering how the whole experience then and thereafter has been photographed and distributed, and continues to survive in an infinite number of pixels for eternity. The means of communication being what they are, the extent of the destruction was quickly seen, not only in Nepal but the world over.

Considering how rare cameras were in those days, the grainy, black-and-white images capturing scenes in the aftermath of the 1934 earthquake were, unfortunately, limited to Kathmandu. Photographic documentation of its impacts outside of the Valley is almost non-existent. That was certainly not the case in 2015. The ubiquity of the digital camera, and more so of the smartphone, ensured that everything that could be was documented by professionals and amateurs alike.

* * *

Due to Nepal's location on a geological fault zone, various earthquake scenarios had been envisaged over the years, with almost all of them focusing on the Kathmandu Valley. With good reason, too, since the unplanned growth over the decades had made it ripe for what experts believed would include a near-flattening of the Valley's urban cores. That the capital would be cut off from the rest of the world with the few roads leading to it blocked by landslides, and the one runway at the airport damaged was the situation imagined. Thankfully, despite the gutting of much of the Valley's cultural heritage, the kind of devastation prophesied did not come to pass.

In fact, in marked contrast to the dire warnings the death toll was surprisingly low. While this is not to downplay the human suffering caused by the earthquake it was sheer luck that the calamity struck around noon of a late-spring day when most people were out and about instead of within the confines of their homes. Another piece of luck was that it was a Saturday, the weekly holiday in Nepal, and students were not in school. That perhaps spared hundreds if not thousands more casualties given that some 9000 schools crumbled to earth or were damaged badly.

For all it wrought, the earthquake did not actually consume the entire nation. Damages were extensive in the Kathmandu Valley but fewer than half of the country's 75 districts were affected and of those only 14 seriously. The losses were not in any way insubstantial. Nearly 9000 people were killed and hundreds of thousands of private houses, public buildings, government offices, and historic and archaeological sites either destroyed or damaged. The total losses to Nepal due to the earthquake was estimated at USD 7 billion.

Immediately termed a national disaster, the earthquake also had a different kind of impact on the country. The unexpected catastrophe struck at a crucial period in Nepal's political history with a nearly decade-long political transition threatening to drag on indefinitely. Despite a constituent assembly having been elected two times, polarisation on the question of federal boundaries remained strong and passions remained unabated on all sides. Chances of a constitution through mutual compromise were proving elusive. The earthquake and the scale of devastation it caused gave pause to all forces—political and extra-political—that were threatening agitations against anything that went against the mutually exclusive demands all round. It was under such circumstances that the leadership of the major political parties struck a deal in June 2015 which considerably watered down many of the progressive provisions expected to be included in the new constitution. And in less than four months, a 'fast-tracked' constitution that further diluted the provisions of the June pact was presented as a fait accompli to the people of Nepal to the accompaniment of violent protests against it rocking parts of the country.

The political stalemate also spilt over to post-earthquake reconstruction. In a quick call for assistance, Nepal hosted a conference two months after the earthquake and secured pledges worth nearly USD 5 billion from a host of countries and organisations. But despite the urgency, it took eight months before the government could set up the oversight body for the process, the National Reconstruction Authority. True to style, political parties squabbled over it and thus delayed its establishment at a time it was needed most.

* * *

Far from that world of politics, the photographs in this book tell a story of both human resilience and perseverance. Amid the visually arresting images of the destruction, here are ordinary people going about their business—whether as suffering victims or as entrepreneurial workers. The photographs have been further enriched by a group of Nepali writers who have provided somewhat personal and visceral accounts of how they view the earthquake or were affected by it.

It needs pointing out here that not all the photographs have been captioned since getting the information required would have required Kevin to intrude into people's lives at a time of hardship and even tragedy, or otherwise disrupting livelihoods, howsoever fleetingly. Instead, there is the bonus of the 'Photographer's journal' at the end of the book, a section that should be read in conjunction with poring over the images to understand the circumstances under which they were taken. The stories these photographs tell were worth capturing, and Kevin knows how to do that better than most.

Narratives of a disaster

— Kevin Bubriski

I did not arrive in Kathmandu until early June 2015, more than four weeks after the 25 April earthquake. It was a journey of choice back to Nepal and would have been weeks earlier except for my teaching responsibilities as a college professor of photography and documentary studies.

I had last been in Kathmandu in January for the launch of my book, *Nepal*: 1975-2011, a 36-year retrospective of images. My trip to Nepal in June was to see for myself old friends of years ago and to bring my camera to witness what had changed of a place and people familiar to me over the past 40 years.

The initial trauma of the disaster had been over by the time I arrived. The rescue efforts had happened weeks earlier and even relief had slowed to a trickle of what had taken place in the aftermath. The many young Nepali men and women such as those at the Yellow House who had coordinated the relief through crowd-sourcing funds and also managed the transport of food, water, tarps and essential relief supplies had for the most part returned to their usual routines of life and work.

A first impression of mine was the large number of pallets of relief supplies under nylon strapping at the airport. It was bewildering to see how all the supplies could be sitting there on the tarmac so many weeks after the disaster. The second impression was that everything appeared rather normal on the drive from the airport to Patan. Nothing shocking and really no evidence of the tragedy, the trauma, and the devastation. Once in Patan, there were the closed streets, splints and wood struts supporting buildings shaken and leaning from the quake.

I wanted to take my time over the month in Nepal to meet with people in the Kathmandu Valley as well as in the hill regions. I thought I would tag along on relief operations, but actually ended up spending $100/day to hire a driver and a four-wheel drive to get out of Kathmandu and up into parts of Sindhupalchowk. I usually travel on local transportation but due to landslides from destabilisation of roads and mountainsides following the quakes and the monsoon rains, for my first time ever in Nepal I felt like I was moving around like a government or NGO official.

In Nepal, I have never really been on any sort of news cycle or deadline. I work as a non-fiction fine art photographer, a documentary storyteller, combining images and the stories

I learn from Nepali and expat friends living in Nepal. I used to say in the 1980s, when I travelled for many months at a time over a six-year period, that my only deadlines were the expiration dates on my boxes of large-format sheet film.

Even a month of time flashed by far too quickly. Due to the precarious road conditions and the limited time, I never got out to the village areas I knew well in Nuwakot, Rasuwa and Gorkha districts. I did cover several places in the Kathmandu Valley, but again never had enough days to spend in Bhaktapur, Sankhu, Khokana or Bungamati to do more than visit a few times at best, opportunistically make my photographs, and gather enough information for extended captions. I became particularly interested in photographing the 'building breakers'—who with a minimum of means, tools and pay often lived in the buildings they were deconstructing. My regret is that I did not spend more time with them to more fully tell their stories.

Upon my return from the month in Nepal, I was surprised to find that the New York Times Lens Blog and *New York Review of Books* wanted my images to update their readership on events in Nepal. For two months, since right after their initial coverage of the quake, the *New York Times* had not really covered Nepal until the publication of my work in early August. This is indicative of how news cycles work and how they are dependent on both the public's attention for an established story and appetite for fresh breaking news events.

Now, seven years later, the photographs from the aftermath of the 2015 earthquakes have lost that visceral immediacy as news, but the photographs are reminders of an historical event that tore apart the ancient and beautiful cultural integrity of Nepal. One now sees new, tall concrete, steel and glass residential and commercial buildings where the red earthen brick and wood-carved buildings once stood. As reminders of what was lost, the photographs now have their own independent lives as historical artefacts.

Acknowledgements

My heartfelt thanks to my wife, Laura, and our children, Tara and Ryan, for their love and steadfast support.

I am indebted to Nayantara Gurung Kakshapati, Bhushan Shilpakar, Nishant Shilpakar, Sagar Chhetri, Shikhar Bhattarai, Thomas Pouppez and Shanti Lama at Photo.Circle Kathmandu for their friendship, photographic insight and logistical support. Photographer Chemi Dorje Lama and driver Shiva Rai made travel through Sindhupalchowk and Dolakha districts rewarding and safe. Thanks to the medical and nursing staff at the CIWEC Hospital and Travel Medicine Center for their excellent help. Journalist Donatella Lorch was my charismatic guide on our walk through the devastation of Khokana village. I've known dear friends Bharat Jung Upreti and his wife, Bimala, for four decades. I welcomed their son, Alok, to accompany and assist me at some of the demolition sites.

Thanks to James Estrin at the *New York Times*, Lucy McKeon at *The New York Review of Books Daily*, and Sienna Craig and Mark Turin at HIMALAYA: *Journal of the Association for Nepal and Himalayan Studies*, who in 2015 published selections of the photographs in this book.

Deepak Thapa's fine vision as writer, editor and publisher was critical for this Himal Books production. Norbo Lama has been wonderful to work with on the design of the book. His quick eye is balanced with his meditative care in his design work. Manjushree Thapa's enlightening introduction to the book deepens our understanding of the photographs, the earthquake and its aftermath. My sincere thanks to all the writers who generously contributed their insightful essays. The book is greatly enriched by their stories of personal experiences and observations.

—Kevin Bubriski

KATHMANDU

Above: The wooden struts and buttresses can hopefully hold up the massive walls of the Hanuman Dhoka Darbar.

Left: Partial facade of the Gaddi Baithak aspect of the Hanuman Dhoka Darbar.

Overleaf: Scaffolding protecting the statue of the deity, Hanuman, at the entrance of the Hanuman Dhoka Darbar.

का दरवार

The site of the iconic Kasthamandap public shelter, Darbar Square, Kathmandu.

Above: The remains of the Maju Deval Temple, Darbar Square, Kathmandu.

Overleaf: The Trailokya Mohan Narayan Temple in the main foreground with the Maju Deval Temple to the left and the white facade of the Gaddi Baithak in the background.

DANGER ZONE
DANGER ZONE
DANGER ZONE
DANGER ZONE

A long time coming

— Prawin Adhikari

Early one Thursday morning in May 2015, at Mike's Breakfast, a restaurant with a garden under leafy trees in a quiet neighbourhood in Baluwatar, Kathmandu, the Rasuwa Relief team meets to coordinate the deployment of a relief mission. A peacock crows in its enclosure in the garden while a popular actress, a prominent cardiac surgeon, and a famous businessman drink their morning coffee. Austin Lord, a Fulbright research scholar from Maine in the US, here studying the 'Nepalese hydroscape', worries his cropped red beard while his new bride, Sneha Moktan, teases him about it. Each of the half dozen volunteers present make frenetic calls to procure the tarpaulins and food needed by the thousands of people living without shelter or access to food. Three trucks are on their way, but it isn't clear whether they would come to Mike's Breakfast, or if they have chosen another point of departure.

At 9 am on 25 April, Austin and Sneha had left the village of Langtang, at an altitude of 3,400 metres, a day ahead of schedule to return to Kathmandu where, after a week, they were to get married. The couple-to-be had managed to squeeze in a short trek before their big day. As they sat down for lunch at Ghodatabela, a village a few hours down the trail to Kathmandu, the mountains rumbled and gave them just enough time to rush out. Houses collapsed within minutes. Boulders the size of houses hurtled down sheer cliffs on all sides. In Langtang, a landslide erased the settlement from existence. The valley emptied, abandoned within days by the survivors when there was no hope of rescuing those who didn't escape the deluge of debris.

Austin and his friends, who had long and abiding ties with the Langtang Valley, started Rasuwa Relief, a spontaneous volunteer effort to bring basic shelter and food to people who had lost their homes. There is no exact count of such volunteer groups like Rasuwa Relief. But they have been highly effective at reaching and supporting survivors in the most-affected areas of the mid-hills and mountains of Nepal.

Himalayan Disaster Relief, a collective operating out of Yellow House, a bed-and-breakfast in better times, has conducted more than 200 relief missions. The CYF Collective at Base Camp—a combination of a small NGO focused on children and youth, and friends who patronised an artsy bar—has moved beyond food and basic shelter to rebuilding the village of Kalinjor in Kavrepalanchowk District, some 60 km to the southeast of Kathmandu. Whereas the government has lumbered, such groups work with electric urgency for the simple reason that communities affected by the earthquakes have limited access to food and shelter. And, the monsoon is coming.

Three trucks carrying 192 sacks of rice and other essentials are scheduled to leave for Shantibazaar, northwest of Kathmandu on the border between two hard-hit districts—Nuwakot and Rasuwa. The food is earmarked for over 600 displaced families from Dandagaon and Thulogaun villages in Rasuwa who have gathered some distance to the south of Shantibazaar, in and around a camp for Chinese contractors working on a hydropower project on the Trishuli River. At noon, after an hour of rearranging the sacks of rice and salt to evenly distribute the load between the three trucks, the convoy is ready to leave.

The drivers want to take the shortest route—passing through their village, through the jungles of the Shivapuri National Park. Relief trucks have been looted by mobs of citizens angry at the government's slow and ineffectual response to their situation. Politicians have been accused of rerouting relief convoys to distribute goods to their constituents, even at the cost of communities facing starvation. 'We'll take the safest route,' says Bikram Karki, a 28-year-old adventurer and trekking guide who has worked with Austin as a translator and research assistant. 'We don't want to risk anything,' he says, even though the drivers complain.

Bikram is quick to laugh and has an instinct for ingratiating himself with strangers. His wife wants him home by the evening. He has seen his share of troubles around distributing food and tarps, so he is prepared for a long slog: a few hours to gather information, another few hours of negotiating with the locals. Quite possibly, the food will be distributed the next morning.

The convoy travels west from Kathmandu before turning north along the Trishuli towards Shantibazaar. Along the highway, houses appear normal until a closer look reveals how they have fractured. Through open doors and windows appear the mess of collapsed roofs and the absurdity of wooden staircases lying on their side. Baby mangoes the size of a child's thumb cluster above dark shades in which people fan themselves. During the day, unlike in tent cities in Kathmandu, very few people sleep under the tarps slung between short trees or from bamboo frames. They lean against the fruit trees in their yards, unable to shake off the stasis that has transfixed them. There is terror in their eyes from always waiting for the next tremor. Men fold their shirts up to expose the soft flesh of the belly and women pull their saris up to their shins, digging the soles of their feet into the dirt for some cool. But they do not venture far from the cracked shells of their homes, from where they still hope to salvage whatever they can, or which they guard against theft.

The police along the highway have been tasked with providing protection to groups like Rasuwa Relief. At Galchhi, where the road forks north, the trucks are inspected for paperwork and permits. A policeman appropriates a fluorescent lantern with in-built solar battery donated by the students of Malpi International School in Kathmandu for people living under tarps in Rasuwa. He gives it to a woman in a teashop across the police booth.

A few kilometres on, the police stop the trucks again. An officer climbs on. When he disembarks to make way for a colleague, he carries away another lantern. The police stop the trucks after every few kilometres, but they don't take any more lanterns.

Since Shantibazaar is on the Nuwakot side of the border between two districts, the Chief District Officer of Nuwakot needs to give permission to distribute goods. But the displaced are mostly from Rasuwa, and therefore not his responsibility. A quarter of an hour passes before a message is scrawled on the back of a letter, giving the trucks passage to Shantibazaar. The writing is illegible and the language bureaucratically opaque; the paper never comes into play again.

The Trishuli is usually glassy and green in May. But it roils murkily that day, muddied by landslides upstream. If calamity weren't still frightening the people along the river, they would be out tending to the young maize crop or watching the fat, golden sheen over fields of ripe wheat. There would be fish to catch in the river. But the masses are huddled now. Sometimes the rubble of a house has become a neat pile of salvaged wood and stone. Mostly, hollow-eyed families sit immobile by the roadside and watch everything with an unnerving intensity.

At 4 pm in the afternoon, after driving 130 kilometres, the trucks stop at the gates of the Chinese camp, where contingents of the Nepal Army and the Armed Police Force are stationed. It is a large compound, with a few hundred apartments, and hundreds more under construction. The yard looks clean because it is covered with smooth pebbles from the river. But it is, in fact, littered with butts of Yuxi cigarettes, a Chinese brand not available in Nepal. Very few of the Chinese engineers and workers remain—most evacuated when the quakes began. A soldier shuts the gates behind the relief trucks to keep out people who aren't taking shelter there. These are people from the immediate villages in Nuwakot—Shantibazaar, Archale and Salimbhitta. They are not the intended recipients of the bags of rice, dal, salt and oil.

'Do you have enough for everyone?' an off-duty soldier asks. A few hours earlier, two trucks had returned with their cargo of relief material because they had started a brawl among the locals, he says. The policemen asks for a list of the goods Rasuwa Relief has brought, and whisks away Ram Krishna Bhatta, the local volunteer who has accompanied the trucks from Kathmandu. They want the decisions about distribution to be made in the most secluded apartment—which they occupy.

Bhatta disagrees. 'The division has to be done where everybody can see us. They will blame us otherwise. They will get angry,' he says.

As he walks away, a policeman pulls aside another volunteer and says, 'Our Inspector has said we desperately need two sacks of rice. We haven't received any aid either. Two sacks of rice, and four litres of cooking oil.' Throughout the afternoon, as Bhatta divides and subdivides the sacks of rice, dal, salt, soy nuggets and half-litre packets of cooking oil, a volunteer has to remind Bhatta to discount the rice and oil that has disappeared from the official count.

Three different ethnic groups—Tamangs, Gurungs and Bahuns—are being asked to share the relief material. But the Tamangs and Gurungs are suspicious of the Bahuns: they speak the language of the establishment, and they have long known the American who sent the goods. Before rising voices can boil over into chaos, before harsh words are spoken and neighbourly bonds broken, Bhatta pulls to his side the angry young Tamang man who claims that relief missions have been neglecting the small cluster of houses around his home. 'Sit here,' Bhatta says. 'Tell me how this should be done.'

Representatives from each village ward huddle around Bhatta, who is occasionally heard asking, 'Does this work for everybody?' Behind them, neat piles of rice and dal sacks grow on the gravel compound of the Chinese camp as men from Archale empty the trucks and divide everything according to the numbers agreed upon by the representatives. Young men take photos with 30-kg sacks of rice on their backs.

Two villages receive 72 sacks of rice each, along with other provisions in equal proportion. Every village gets a bit of everything—some get more rice, some have to make do with more bags of soy nuggets. The murmurs of debate and dissent settle into grunts of approval and acceptance, and people living in the camp and from nearby villages discuss rebuilding.

As the rain clouds hover with the closing of the day, villagers stir once more into restlessness. An old grandmother wearing flat plates of gold bigger than her ears stands from her spot by the open door of a prefab apartment. Everybody sleeps by the door—the quicker to run into the open if the earth trembles again. A dusk made gloomy by an imminent storm brings back old fears of landslides, of the cold, of damp nights spent under an open sky, and makes people think of shelter for the night, for every night to come.

Two men in their 50s try to shake the prefab apartments of metal and plastic panels. They talk animatedly of the moment the first big quake brought down their houses. Then they worry about the monsoon. They are Bahun farmers, possibly just as illiterate as their Gurung and Tamang neighbours, with dark soil under their toes and a look of bewilderment. Steel braces secure the panels. One of the men, with thick knuckles and bulging veins, asks, 'How long do these houses last?'

'Steel doesn't rot. Plastic doesn't rot,' his friend knocks on the panels. 'This is the kind of houses we have to build.'

'If it lasts another 15 years, another 20 years, I can die satisfied. At least, I will have left something for my son,' the first farmer says. 'For his sons, he will have to build.'

SWAYAMBHUNATH

Members of the Nepali Army pitched in on the demolition of unsafe shops and houses on top of the Swayambhunath hill. Fortunately the ancient Stupa, seen here in the background, suffered no major damage.

Sanil Buddhacharya, 13 years old and studying in Grade 8, has been sharing this small sleeping and living space with his mother, Subrana, and father, Swayambunath Ratna, since the earthquake eight weeks earlier. Twenty-seven other families from the Swayambunath locality were internally displaced people living under tarps among the ruins there.

By mid-June there was little left of the Karma Raja Maha Vihara monastery on Swayambunath. A major seat of the Karmapa's in Nepal, the monastery had been badly damaged and was being taken down under the supervision of UNESCO and the Karmapa ShamarNag Committee.

Soldiers of the Nepali Army assisted in the demolition of unsafe shops and houses on top of the Swayambunath hill. The ancient stupa stands in the background without major damage.

Nepali resilience

— Anil Chitrakar

Nepal is beautiful because we have earthquakes. The breath-taking landscape that the whole world aspires to visit and experience at least once in their lifetime is the result of faults, energy stored and released over millions of years. We must all know this, we must recognise this, and we must live with this reality or pay a very heavy price. When the earth began to move below us on 25 April 2015, the first reaction was to hold on to something for support. It then hit me that everything was moving. I then noticed the flowerpots were rolling like a spinning top and falling over on the side. The brass oil lap hanging on a chain from our rooftop began to sway and remained our principal indicator of numerous aftershocks for months to come. Looking towards the old parts of Kathmandu, all we could see was a cloud of dust.

Nepal rests on Saturday

Thousands of classrooms were damaged during the 2015 earthquake. It is hard to imagine the death toll if they had been full of children. Seven hundred thousand houses were damaged. Luckily, many people like me were in the garden or the fields, weeding the maize crop. Seven hundred-plus monuments fell during the earthquake and its being a Saturday these were crowded with people. Some were donating blood. The earth began to move during the day and not at night, another factor contributing to fewer than 10,000 deaths. The day of Saturn, or *Sanischar*, may mean different things to different people but for many Nepalis, Saturday is a lucky day. Next time, we may not as lucky; hence, we need to be prepared. The energy is building under us.

Nepal united

Politics, monarchy, the constitution-writing process, the armed conflict, had all contributed to dividing Nepali society. The earthquake in a strange way united Nepal like nothing else could. Homes, kitchens, compounds all opened up for the needy. People cooked in common kitchens and slept in makeshift tents made from bamboo and plastic sheets. Children and the elderly were looked after by everyone. People spoke kindly to each other and many used their bare hands to dig out those buried under the rubble. People ferried the injured to hospitals. The telecom and power utilities brought back services quickly and the radio kept people informed 24 hours a day. In a strange way, Nepalis felt good in the days after the earthquake.

Nepal has many friends

The restored temples at Patan Durbar Square, the rebuilt Durbar High School at the heart of Kathmandu city, and all the families that received support to rebuild their homes are proof that Nepal has many friends around the world who wish to help and support in times of need. All the rescue teams from all over the globe with their dogs and equipment saved many lives. Nepalis and friends of Nepal mobilised and helped in whatever way they could. Some took advantage and shamelessly stole as well. A few weeks after the earthquake, the whole world pledged billions to the rebuilding process. At Boudha, people donated money and gold and the reconstruction began. Mani Lama took a picture of the process every day and has published a book. The Indian blockade of Nepal sent a cold chill down our spine... but we remained determined.

Nepalis understand markets

Nepal avoided shortages and black markets in 2015 by letting the market do what it is supposed to do. The business community stepped up by donating food and other essential products with the best of intentions. It took convincing to understand that earthquakes damage houses and classrooms, temples and structures, but it does not destroy retail and wholesale distribution systems. The sooner these are functional, the better-off we would all be. Free goods can and will distort the market and can have long-term consequences if small and large businesses collapse and have to start all over again.

Rise of artisans

The reason Kathmandu still has so many artisan families is due to the need for their knowledge and skills in recovering our World Heritage Monuments every hundred years. This time was no exception. When heritage falls, the artist rises. In the 1990s, at the Nepal Heritage Society with support from UNESCO and the Kathmandu Valley Preservation Trust, we had prepared a manual called 'Preparedness, Response and Recovery' of heritage sites with earthquakes, theft and fires in mind. This paid off in 2015 when artists stepped up to the task of first removing and safely storing precious objects, then repairing and installing them back where they belong. Years on, these artists can be seen and heard all over the Valley and outside doing what they are really good at. Tourism has helped these artists by creating a market for replicas.

We know how to make chlorine

At the office of the Environmental Camps for Conservation Awareness (ECCA), we had been making and distributing active chlorine for many years. One litre of chlorine can treat 4,000 litres of water and make it safe for drinking. On the day of the earthquake, many people moved out of their homes and set up camp in the nearest open spaces. They collected water from taps, ponds, water spouts, and tanker services. It was starting to get warm and if people became sick, we could have an epidemic in our hands. The entire ECCA office was emptied and the whole building became a chlorine factory. We had also made sure villages could produce their own chlorine using a solar panel-powered unit. The chlorine saved many more lives in the days and months ahead.

Save the seeds

Nepali traditional wisdom says that 'in a mob you save yourself, but in a disaster you save the seeds'. After the 2015 earthquake, Nepal's airport, warehouses trucks and just about any vehicle you came across were full of 'free' food for distribution. This 'free' food and the crowding out it could create for the locally grown food could have been one of the biggest challenges in managing the recovery process. Nothing can compete with 'free'. The pre-monsoon rains had just begun and if people did not plant the precious seeds, Nepal's real disaster would begin in about six months when free food would be gone and there was nothing growing in the fields. Nepal responded to its traditional wisdom, got the seeds out, planted them, transplanted them, and all was well for the festivals in October.

Build back better

When the 2015 earthquake shook Nepal, we were still struggling to complete the new constitution and people were really tired of the politicians and the process that had dragged on for eight years. We did not have anyone elected to our local ward and municipal offices. This absence became a huge problem in collecting data and managing the resources, and in the distribution of materials and money. The government announced the formation of the Reconstruction Authority and the slow pace of reconstruction began. Years later, many people were still filing the paperwork, applying for assistance, and rebuilding their homes and lives back. Grants continue to flow from around the globe. In the meantime, the de-population of villages is speeding up.

Death gives meaning to life

One of the best things that happened for Nepal and the tourism sector was the release of the Marvel movie *Dr Strange*. The film helped put Kathmandu and Nepal back on the map of tourist destinations. There is a line from the movie that goes, 'Only death gives meaning to life.' It is the acceptance of death as part of the cycle of life that helps us to live better. Across Nepal we now meet thousands of Nepalis with their families, having a good time, spending their hard-earned money. Many have told me the earthquake really changed the way they looked at life and death. Let us remember the earthquake and the dead and learn to live well in this beautiful country.

NAYA BAZAAR

Shiva Maya Upreti sits beside the photograph of her younger sister, Krishna Thapa, known to all as 'Kanchhi Didi'. Kanchhi Didi passed away 13 days earlier due to pneumonia, exposure and complications from living and sleeping outside of the family's cement house in Naya Bazaar. She stayed outdoors as she was fearful of being in the house during the strong aftershocks. She is survived by her husband, three children, three brothers and her sisters, Shiva Maya and Hari Maya.

SPORT

Facing page: This young man and 11 co-workers from the same village outside of Udaypur are part of the demolition crew taking down a four-storeyed concrete house in Naya Bazaar. They make Rs 700/ day (around $6) while the demolition contractor provides them with food.

Preceding spread: At mid-morning, labourers take a break from their heavy work tearing down buildings. These four young men came to Kathmandu looking for work after the earthquakes. They lived in the building they were tearing down with eight other friends from villages in the southeastern district of Udayapur. These workers, numbering in the several thousand across Kathmandu and other affected areas, were paid by the owners of the properties they demolished.

Facing page: Sita Karki and Sunita Pariyar work on a building demolition team with six men. Sita and Sunita's job is to carry *doko* basket loads of brick and debris down the debris-strewn stairs and out of the building.

How to build a home that lasts

— Muna Gurung

In Arthar, Parbat, my Baba, who is 75, runs up and down trails like a little boy. I can hardly keep up with him. He will point at a large bush and say, 'There should be a path that cuts through here.' And sure enough, there will be. In Arthar, I don't worry that Baba will slip and fall, I don't ask him to *be careful* like I always do in Kathmandu. There, the ground holds his feet as he sprints, the air lifts him by his arms, they recognise the boy he once used to be in that village, the boy he still becomes when he goes there. Baba knows Arthar and Arthar knows him.

/

In the 27 years he spent in Singapore as a *lahure*, the city never became home for Baba. Singapore gave him his first pair of shoes: they were green canvas boots with soft rubber soles that ate at his unaccustomed feet, leaving tender blisters. Singapore also gave him his first transistor radio, his first watch, and me, his only daughter. But it never gave him a home. We lived in a small, enclosed part of the city where all the 'Gurkhas' lived, we called it our *camp*. As a family, we moved from one flat to another depending on how well Baba performed as a *lahure*.

'They were not mine,' he says about the flats. By which he means they were just rooms built to house bodies, they were not built with him in mind.

When I ask Baba what home is for him, he laughs and says, 'What kind of a stupid question is that?' For him it is simple. Home is a house that belongs to you, that you build on a property that you paid for, that you own.

'That means I don't yet have a home,' I say.

He smiles because he sees that the answer is not so simple anymore. 'This is your home,' he says to me, pointing at the ceiling of the Kathmandu house we live in. 'Because home is a place that loves you. This house loves you.'

/

'We used to buy nice pots and pans, but we never used them,' Aama tells me. 'We wanted to save them for a home we would one day build in Nepal.'

It is amazing to think about how hard humans must actively try to *not* build a home wherever they are. To decide *not* to root, not here. 'Singapore was always temporary,' she says.

Home, then, is permanent.

/

In 1994, Baba retires, and we move back to Nepal where he builds us a home in Kathmandu. He uses the best quality marble, wood, brick, iron and cement. For the next two years, he goes to the site every day to inspect every inch of the place. 'It is built to last,' all the relatives laud him.

Years later, when I am half his age, Baba tells me a secret: He hates the city. The plan was always to build a house in Arthar. Buy a horse, ride around the village on that horse. Buy some buffaloes, a crowd of chickens, and spend his days in the fields.

'But the village was emptying, and I stopped recognising it,' he says with a faraway look. 'And your Aama was always sick...it only made sense to be living closer to good hospitals and doctors in the city.'

Home then, is always in the future, a dream and ultimately a compromise.

/

A boy once made a home for me on three pieces of paper. On the first one, he drew the front of the house, in it he planted a small garden of marigold and daisy. On the second sheet of paper, he drew a map of the layout inside, two whole floors of it. On the third, he made an open-air bathroom for me. It was semi-circle in shape, and I could look at the bamboo grove as I showered.

/

When I first visited Nepal as a little girl, I was shocked at all the single, stand-alone houses. 'Nepal is a rich country!' I would report back to my friends in Singapore. The only homes I had known then were small apartments like ours where limbs, love and arguments packed tightly together. I never knew houses were allowed to be big.

In those days, home meant the sound of the fish tank in our living room; the breeze that blew in the hallway of our apartment where Aama and I laid out Styrofoam mats for afternoon naps; the evening walks to the cemetery; the softness of the sun at 5 pm, the sounds of mothers playing badminton, and children playing basketball, hopscotch, rubber band; the sounds of pressure cooker and television sets broadcasting the Gulf War; walks to the dumpster carrying leaky bags of garbage; and the excitement of sleeping on the balcony on hot nights.

Home is in the past, then, a place we continue to leave behind.

/

At 19, like Baba, I pack two suitcases and make my own journey across the world.

When I get sick, I squeeze lemon over my food, 'It will make anything edible,' Aama used to say. One year, someone who loves me enough sends me an audio recording of the market in Lagankhel. It is two minutes long and I listen to it with eyes closed: Rs 30 for eggplant, Rs 120 for apples, Rs 100 for three pairs of socks. I am back in the crowd, shoulder brushing against shoulders, the smell of citrus and rot in the air.

/

In 'Damascus, What Are You Doing to Me', poet Nizar Qabbani describes the letters his mother sent him every spring. He would open the letters in London to find in each, a bunch of tarragon. The British would be confused, they would interrogate him: Is it a talisman? Medicine? A secret code? He writes:

> *I said to them: It's difficult for me to explain...*
> *For tarragon is a language that only the gardens of Sham speak*
> *It is our sacred herb...*
> [...]
> *My mother is a wonderful woman...she loves me greatly...*
> *And whenever she missed me*
> *She would send me a bunch of tarragon...*
> *Because for her, tarragon is the emotional equivalent*
> *To the word: my darling...*

Home is a herb, a medicine, a message from very far away.

/

One April morning, I wake up to an email with a photo of Aama in the garden. She is seated on the bare ground, prayer beads in hands, eyes closed. The dogs are next to her, Baba is behind the camera. The message reads, 'we r fine.'

It doesn't take long for me to find out that while I was sleeping 12,000 miles away, an earthquake has arrived in Kathmandu. The ground is still moving, sometimes in waves, sometimes in circles, and sometimes in jolts. I call home but I can't get through. I wait patiently for photos to surface on social media. When the images trickle in, I see homes uprooted like radishes, their insides exposed. How the geography of the inner lives of families are now so easily cut open, now familiar—there is a plastic-rimmed mirror that hangs on the wall, the flimsy paper calendar flaps against the wind, the wall clock is slanted, its hands don't move, and the toothbrushes stand bewitched in a cup on a sink.

/

Years later, when I visit Manekharka in Sindhupalchowk, I am taken aback by the beauty of large black metamorphic rocks that litter the village. Origin stories are aplenty, but the one I love and hate the most is how Shivaji was frolicking with a bunch of women and when he was caught by Parvati, the women turned into rocks. That is the lot of women.

There is one particular spot where the land dips to create a large pond.

But when I go there again in the winter, I can't seem to find the pond. The water has left, and in its place there is a lush green carpet of grass and rocks, large black rocks. I walk down to what used to be the pond and sit on one of the rocks. What had I thought the bottom of a pond would look like? I am in that bottom of the pond and the air around me still retains the quality of water. Now, I sit in absence. I sit in the ghost of a pond. And only the pond knows why the heart aches.

GONGABU BUS PARK

ME
nal Money Express
ई मित्र छ

D1
POL 01
28 APR
E

Kumar Rana works on demolition. He is the owner of the Tanahu Laxmi Magar Guest House in Gongabu.

-02
DOL-01
2MAY

F-02
POL-01
2MAY
GSCF

Memories of a disaster

— Tsering D Gurung

Erasing unpleasant memories at one's own will is an unattainable feat even for the most gifted among us. We would neither know trauma nor psychological distress had it been otherwise.

It's been many years since the Gorkha earthquake ravaged the nation; killing over 9,000 people, injuring thousands more, and destroying hundreds of thousands of homes in its wake, but for those of us who felt the deadly tremors that fateful April morning, the memories from the day remain vivid.

Back then, I was working as a journalist at a weekly newspaper based in Kathmandu. My colleagues and I had gone on a weekend retreat to a resort located on the Valley's outskirts and were preparing to check out when the earthquake struck at 11:56 am.

Overcome by shock, fear and anxiety, we watched in silence as the city in front of us started to get engulfed in a sea of orange and brown dust. All of us tried to nervously dial our families, but the lines were all dead. There was nothing to be done except wait for the earth to calm down.

'This is the big one,' my editor at the time noted, looking forlornly ahead into the Valley.

For years, we [the paper I worked for] had been covering earthquake-related stories, analysing how unprepared the country was for the next major quake; assessing the extent of damage and human casualties and chaos such a disaster would result in; and educating readers about the safety measures to be adopted in order to mitigate the impacts.

Our stories reminded us and the readers that in Nepal, it was always a question of when, not if, a major earthquake would strike. After all, it had been over 80 years since the country had last experienced a big quake, and experts had been warning that our country, located on an active seismic zone, was long due for another massive one.

But it was not until the morning of 25 April 2015 when the violent tremors woke us up from our deep slumber that Nepalis truly understood the perils of living in an unprepared quake zone and realised just how woefully unprepared we had been despite past warnings.

Even the most informed of us had failed in our preparations.

While at my job, I had edited articles recommending readers to prepare 'go bags' and instal seismometers in their homes and offices. I had failed to follow that very simple advice myself. The paper may have published multiple stories highlighting the importance of following safe building codes, but its own office continued to be housed in a complex with no emergency exit. And, perhaps most importantly, as a newsroom, we had failed to devise a contingency plan in the event a major disaster hit the country.

How would a newsroom function; how would reporters commute and communicate amid the infrastructure crash; how much risk should one consider taking during disaster coverage? Will the office be held liable for an employee in case of an unfortunate incident? These were questions we had not pondered upon during editorial meetings.

For many of us in the newsroom, it was our first time ever covering a catastrophe of such a scale. Despite our inexperience and lack of training, we were eager to do our jobs, to continue telling stories in order to inform and educate, to do our bit to document this historical event and its aftermath, and to perform one of the core functions of journalism: to hold power to account.

Journalism is a gruelling profession, even on an ordinary day when your routine isn't regularly interrupted by aftershocks and emergency evacuations. Unlike so many other jobs, the work of a journalist doesn't end when the clock strikes 5. The concept of public holidays is implausible. And when a disaster comes knocking, you don't run towards safety, but often to the front lines.

It is, however, also one of the most rewarding. One of my favourite stories from those days is about the rescue of Sonish Awal, dubbed the 'miracle baby' by the press. Sonish, then a four-month-old baby, was rescued alive after being buried under the rubble for 22 hours in Bhaktapur, and the photograph of the moment captured by Amut Thapa became one of the most iconic photos of the quake. Sonish's story lifted up many a spirit during that desperate period.

But while there were stories of hope and happy endings, they were easily outnumbered by those of death, loss and suffering. Witnessing buildings crumble to dust, hearing people wail as they recovered bodies of loved ones, and walking through rubble with the stench of death pervading the air, are not experiences one can easily erase from the mind.

To this day, I still think about the elderly couple in Sankhu who had lost their home and were frantically attempting to salvage their belongings when I met them a few days after the quake. I think about the Dalit family living in Barpak, the epicentre of the quake, who had been cut off from much of the relief supplies because they were isolated from the rest of the village. I think about the people of Langtang, many of whom had lost their entire families, and had been wracked by guilt about being the sole survivors.

Journalism schools and newsrooms train you to spot news, write effective leads, and conduct successful interviews. But no one really talks about how to address trauma suffered as a result of your work or the toll it takes on the mental health of reporters who continue to do their jobs while being personally impacted by disasters.

Even now, I shudder whenever there's a slight tremble near me. On most nights I go to bed musing over my escape plan—just in case another big one strikes—before the day's weariness overpowers my anxiety, and I drift off to sleep.

I think about how difficult it would be to drag along my ageing aunt, now in her 70s and suffering from chronic knee problems and who isn't exactly what one would describe as being light.

I think about whether it would even be possible for me to attempt to rescue my dog which stays on the floor above me.

I think about our house, the place that has been my home ever since I developed an ability to create memories. Like my aunt, the house has aged, and is not exactly in the best of conditions. Just like her, I doubt it'll survive the next big quake.

When the earthquake struck, we felt a bag of emotions, raging anger towards inept officials, overwhelming sadness at the sight of fallen cultural monuments, and deep regrets about not being better prepared as a society; all have subsided with the passage of time. Many of us promised ourselves that we'd do better so that the next time a crisis came knocking at our doors, we'd be prepared.

But as I walked past hastily plastered apartment buildings, read about the government's plan to reconstruct the fallen Dharahara Tower, and heard careless remarks from our officials in the face of a looming health crisis, I realised those learnings were going to be ephemeral.

BOUDHA

Gopini Budhatoki of Dolakha district is 40 years old and lost her home and all belongings in the earthquake. She and thousands of other internally displaced people lived in makeshift tarp shelters as squatters for nearly a year in an empty lot between Mahankhal and Boudha.

Overleaf: Boudhanath Stupa suffered extensive damage during the earthquake.

PREMIUM QUALITY RICE

MUNNI
MUNNI

Biraj Moktan from Dolakha district is 11 years old and studies in Grade 6. After losing everything in the earthquake his family came to Kathmandu as internally displaced people.

Ruptures and revelations

— Niranjan Kunwar

I received this invitation to reflect on the Nepal Earthquake 2015 the day I returned from Manekharka, a village in Sindhupalchowk district hit hard by the April and May quakes. I had travelled with a group of art educators as part of the initiative, 'ArtWorks Sangai Khelaun', launched weeks after the devastating quakes. In the preceding years, apart from Sindhupalchowk, the ArtWorks team has worked with marginalised children and teachers in remote regions like Ghyachhok in Gorkha and Gatlang and Larchhyang in Rasuwa, which were also affected by the natural disaster.

During this recent visit, after completing our day's work at the local government school, some of us were walking down a hillock. At one point, I paused to examine the settlement spread out below. Undulating patches of multi-coloured fields surrounded clusters of buildings. Shades of soil were chequered by greens and browns. But this naturally pleasing view was punctuated by something quite unnatural—swathes of rude blue imported by the locals during the months and years after the earthquakes. These blues were tin roofs initially meant to be part of temporary shelters, but they have become permanent features of villages like Manekharka and Gatlang that were well known for their stone and slate roofs, unique examples of traditional artistry.

Seeing the bucolic, pristine mountain village dotted with these stark industrial building materials was jarring at a purely aesthetic level, to say the least. But this sight triggered another concept that expanded into my consciousness: the concept of violence. I made hasty initial connections. The Manekharka locals—a majority of whom were the historically oppressed Tamangs—were subjected to the physical violence of the earthquakes. But these people have continued to experience a different kind of violence, the violence caused by socio-political systems, the violence that works at psycho-emotional levels. The Tamangs have been ignored by the state for eons, excluded from mainstream politics and disadvantaged by the state education system which uses Nepali as the primary language of instruction. Tamang children, whose mother tongue is not Nepali, have always struggled. In this context, preserving the aesthetics of their traditional villages and monitoring the use of building materials were not priorities for the government during the frenzied period after the natural disaster.

The earthquakes ruptured the earth's surface. And these ruptures revealed known and unknown facets of our socio-political lives even more clearly. It revealed the slow pace of bureaucratic processes as well as the agony of those affected. It revealed the capacity of common citizens to come together and gather resources in an effort to mend the ruptures, in an attempt to heal together, to build peace. On a personal level, it also revealed aspects of my own nature to myself—what I cared about and what mattered to me.

In the immediate aftermath of the earthquakes, there was a need to provide basic necessities to those who had lost their houses and family members. Development workers as well as volunteers and various professionals went on 'relief missions' to places like Manekharka in vehicles loaded with tarps and tents. Since most people were taking tangible measures, some of us felt a need to provide intangible support. How to comfort distressed young children? How to encourage adults who appeared dejected and hopeless? The ArtWorks programme was conceptualised with these ideas in mind. Artists and entertainers collaborated with educators to design workshops for teachers and schoolchildren. It was an experiment of sorts. We went with the belief that artistic practices are valuable, both for short-term relief and for long-term impacts. Over the years, the programme has evolved and provided opportunities for our own team members to learn and grow.

But on every trip, I feel compelled to reflect on our design and revise our goals. Since we face severe challenges while working with government school teachers in Sindhupalchowk, and it is often easy to feel hopeless, I question our intent and purpose, in terms of the specifics and the general. When we reflect as a group, we often use the term 'disrupt'. By introducing innovative methodologies and materials, we intend to disrupt traditional classroom practices and culture. Since government textbooks do not include the mother tongue of most of the children we encounter, artistic mediums like movement and drawing can provide ways of expression. But this technique can sometimes feel uncomfortable. Were we acting entitled and all-knowing? Were we, unintentionally, being violent? Like the blue tin roofs, were we introducing shiny temporary solutions to systems and mechanisms that had functioned, in one way or another, for decades without external interventions? That is why we have to be careful and thoughtful while planning and implementing. In other words, while disrupting, we have to be collaborative and convincing. We have to rely on styles and structures rooted in peacebuilding.

Since we live in a corrupted society mired by an absence of concern for those who are different or disadvantaged, we have to, as scholar Maxine Greene writes, 'find ways of creating situations in which persons will choose to engage in cooperative or collective action in order to bring about societal repair'. The earthquakes propelled many of us who grew up sheltered and ignorant in Kathmandu to take action. Hannah Arendt has also written about the need to 'weave a web of human relationships' in order to create a just and peaceful society. The practice of teaching and learning, of cultural exchange, this act of weaving, has an inherent intangibility. And Arendt reassures us that for all its intangibility, this kind of work is no less real than tangible practices like growing food and constructing buildings.

Yet, I get compelled to question our intent and purpose, to re-examine the scope and value of our efforts, to regularly evaluate the effectiveness of the ArtWorks programme. Our workshops with teachers are informed by an interdisciplinary and integrative approach to childhood education. But I wonder whether teachers who have themselves been deprived of a rigorous education and transformative exposure to these ideas will be able to fully grasp our vision. I wonder whether our philanthropic efforts to disrupt a dysfunctional system will cause more confusion than clarity. I turn to Greene's words for solace: 'Where education is concerned, large-scale solutions hold little relevance for situation-specific undertakings. Local knowledge and local coming together ought to counter the tendency toward abstraction, as should a conscious concern for the particular, the everyday, the concrete.'

By participating in the ArtWorks programme, we realise that we have inserted ourselves into the world with 'a particular kind of identity and responsibility, a particular mode of valuing what lies around, and of straining toward what ought to be'.

Since June 2015, our team members have been travelling to remote Nepali hamlets a few times every year. We pack our materials and belongings and hop into a rented vehicle early in the morning. We share rooms, eat whatever is on offer, witness routines, converse with locals, take walks in the evenings. It was triggered by the earthquakes, this decision to leave our comfort zones and venture into the unknown, this attempt to strain towards what ought to be. The violence gave birth to peaceful endeavours, the violence made us come together, compelled us to imagine, design and learn. There are still a lot of unknowns and doubts, but we know we have to continue this engagement. We have to keep weaving this web.

Facing page: Raju Rai of Sotang village in Solukhumbu district is 20 years old. He never had the chance to go to school. He had been in Kathmandu for 20 days working in demolition/ construction at the Sechen monastery where he was paid Rs 1000 /day (around $8).

SANKHU

Thirty-year-old Purnima Tamang from Okladhunga moves debris by wheelbarrow on an upper floor of a building in demolition. She works for Rs 800/day (around $7).

Together we survive

— Smita Magar

Growing up in Jhumlawang, my remote village hidden in the hills of Eastern Rukum, if there is one thing I knew, it was that disasters are meant to be dealt with together.

I was struck by this realisation when our house caught fire in the year 2014. In the middle of the night as the flames engulfed all we possessed, it felt like the whole village had gathered by our side. And, as everyone worked to extinguish the fire, some neighbours even jumped right into the smoke and fire to help save what they could seize from the flames.

From the very next day, groups of people started coming for the *dukha phalne* ritual—a common social practice in villages in our part of the world to get rid of bad omens. When tragedy strikes, neighbours and relatives of the affected family bring them new clothes, milk or curd, titepati and flowers to pray for better days, show support and solidarity, and encourage healing. People from neighbouring villages also walked hours, even days, carrying clothes, grain, vegetables, almost everything needed to sustain life after a disaster. This continued for months. When we started building a new house, at least one member from every household in the village came to help us clear the debris, carry wood, stones and mud needed for the building. The support continued.

As I had been sent away to the capital to continue with my education after the Maoist insurgency had begun, I had not previously experienced such overwhelming community spirit in dealing with disasters. The holistic support that focused on both the physical and the mental and emotional health at the time of a tragedy was unthinkable to me growing up in an urban setting.

That same year as the fire happened, I had volunteered as an English teacher at a high school in the neighbouring village of Sima. That allowed me to actively participate in different events in my own village. It was during these community gatherings and conversations that I came across the phrase *dharma garyau*! ('You did a good deed!'), generally used by elders when someone undertakes a selfless act. I wondered if our traditional community support system is indeed ingrained in our idea of dharma.

* * *

This idea of doing a good deed and fulfilling one's duty towards others was nowhere more palpable to me than when the 2015 earthquake struck. It was raining heavily as the earth shook under our feet, leaving several among us gasping with fear for our lives. In our village, far to the west of the epicentre, we were afraid our houses of mud and stone would not survive the shakes. We thanked our good fortune to learn that there had been no loss of human lives, no injuries, just a few cracks in the old houses. With telephones and other lines of communication cut, we waited anxiously for the news on the radio as we could sense death and devastation would be mounting elsewhere in the country.

Our community organisation, Jhumlawang Village Foundation (JVF-Nepal), sprang into action the next day with an emergency response meeting. The elderly of the village had spoken, '*Yesto bipatma sakdo jati sahayog garnu hamro dharma ho*' (In these hours of tragedy, it is our duty to do as much as we can to help those in need). JVF-Nepal had been working to build Jhumlawang as a model village but limiting its activities to its confines. Now, at a time of national crisis, it decided to form a volunteer team to mobilise resources and provide primary health care in remote villages of the highly affected areas. The very next day, we left for Kathmandu.

On our way to the capital, we were able to access the phone network again. Feeling relieved to find our relatives and friends in the city safe, we were advised by them not to travel there as the Valley was too dangerous with frequent aftershocks. Unbelievably, we learnt that our rented, albeit old and crooked, house in the city was somehow still standing, and that was to become our base for mobilising in days to come.

When we reached Dang that night, we finally got an internet connection and immediately my phone was bombarded with messages of concern from friends abroad. I realised that the abrupt disruption in communication and being cut off from my closest friends had led to alumni from my alma mater, the Asian College of Journalism in India, on a desperate search for my whereabouts through whatever social media channels they could access. I could sense the reverberations of the tragedy and panic having travelled beyond our borders.

Our arrival in Kathmandu the next day was followed by members of our organisation working actively to secure the necessary permissions for mobilising the medical teams, raising funds, and coordinating with national and international partners. JVF-Nepal's international co-ordinator, Kush Budha, legal adviser, Raju Gurung, and friends such as Pradeep Kumar Rai (Byakul Mahila), the poet who wrote Nepal's current national anthem, ran between the Home Ministry and the Health Ministry for permissions. They were left frustrated by the government's sheer lack of unpreparedness, co-ordination, and availability of data for effective mobilisation of organisations.

'We just wanted a list of organisations, where and for what they were currently mobilised to avoid duplication and immediately reach the communities in the remote villages where no one had gone yet,' Kush recalls. 'But even after five days, everything was in utter chaos. We could not even get simple data.' Without information from the government, we relied on our local networks to decide where we were needed most urgently and mobilised accordingly.

On the 2nd of May, our two groups of medical teams with national and international health professionals and rescue volunteers (firemen) left for Gorkha. One team was to head for Laprak. But on the way the Chief District Officer and Nepal Army officials informed us that it was too dangerous to travel to Laprak, and so both teams were diverted to Simjung. Later, we heard that the same night there huge landslide on the way to Laprak had claimed some lives, including those of a rescue team.

On our four hours' walk from Jhyalla to the Sub-Health Post Centre in Simjung, we hardly saw any house standing. People we came across were either going to their village to attend funerals of their loved ones or returning after attending one. At the health camp we found the expected injuries, respiratory problems, fever, diarrhoea and infections. Many cases, however, were related to stress and anxiety—loss of appetite, insomnia, dizziness, headache and exhaustion.

Many of the elderly who had just buried their loved ones were living with 'survivors' guilt'. The image of 68-year-old Kumsingh Gurung, who had lost his son in the disaster, repeatedly beating his chest while silently sobbing still haunts me. 'It is wrong,' he cried, 'for new generations to die while old ones like me survive!'

In no way equipped to help with such psychological trauma, we could only listen quietly to such laments of losses.

When the second earthquake hit the nation on 12 May, we were heading towards Singati in Dolakha in a bus with 28 health professionals and logistics volunteers. We were about to reach Zero Kilo, when tremors were felt even inside the moving bus. From the windows we witnessed a house next to the road turn to debris, electricity poles sway like leaves in the wind, and dust and dirt rise out of the settlement areas.

Once the bus stopped at Zero Kilo, everyone anxiously tried calling their family members to ensure they were safe. It took a few hours to get back on the road and everyone was determined that help was now needed more than ever. We moved towards our destination, picking up the wounded on our way.

In Dolakha, we found we were the first medical team in the area. The devastation was apparent in the huge cracks on the road, landslides of big boulders, light but incessant rain, all of which made for an ominous sight. The Nepali Army was mobilised to clear the roads. It was getting dark when we reached Charikot. The collapsed and wrecked buildings looked like they were right out of a dystopian film—empty, skeleton-like, and devoid of any form of life within. That night, after *dalmot-chiura* as dinner, the team tried sleeping in the bus amidst jolts of aftershocks. We could not go to Singati as planned because of landslides but we provided health services in the villages of Sunkhani and Sudrawati amid hovering helicopters and frequent aftershocks.

* * *

In the month-long mobilisation, we were able to organise six health camps in Gorkha, Dolakha, Kavre, and Dhading districts, providing primary health services to some 1400 earthquake survivors. We were only able to do it because of our local, national and international partners and friends who extended their expertise and financial support. I could almost see the friends, good Samaritans and even anonymous people coming together as a world community, extending their generosity and support in any way they could.

Despite the chaos at the national level, each region we visited had community resilience in the form of neighbours helping each other, youth mobilising resources, or the diaspora sending in support. We ourselves were supported by local organisations in our work. In one such collaboration, JVF-Nepal collaborated with Karma Flights to airlift four serious cases from Simjung to the hospital in Pokhara. Yet, any assistance has its limitations, based on access to power, information, and infrastructural development (road). For instance, in Sundrawati, Dolakha, people from the Thami community came to us on the last day of the three-day-long health camp as they had not received information on time.

Due to such systemic lapses, traditional community support systems played a vital role for the communities to fall back and depend upon for their survival. In the days following the earthquake, much was said about 'Nepali resilience' through the romantic veil of ultra-nationalism. Unfortunately for Nepal, this resilience could not possibly make up for the power tussle, corruption, and mismanagement of resources at the hands of the government.

* * *

As I look at the present, Covid-19 has rendered even the traditional community support system in Jhumlawang redundant. Unlike during the earthquake, the present pandemic does not even let us mourn together as a community. Health guidelines on physical distancing and safety measures have been highly unpopular here, leading to frustration and disappointment among those who are threatened by the health crisis but are yet conservative about their customs. It is very hard to do the 'good deed' when you may get infected or the ones you are trying to help may be passed on infections because of your actions. In this scenario, the same elders who insisted on doing a good deed after the earthquake are wondering if the saying, '*Dharma garne ko karma phutchha*' (Those who do good suffer an ill fate), has some truth to it.

BHAKTAPUR

Sanchabir Ghising, 29, of Gaushwara 2, Darga village, Ramechhap district.

Jit Bahadur Tamang, 51, Kama village, Ramechhap district.

Prakash Tamang, 35, Kama village, Ramechhap district.

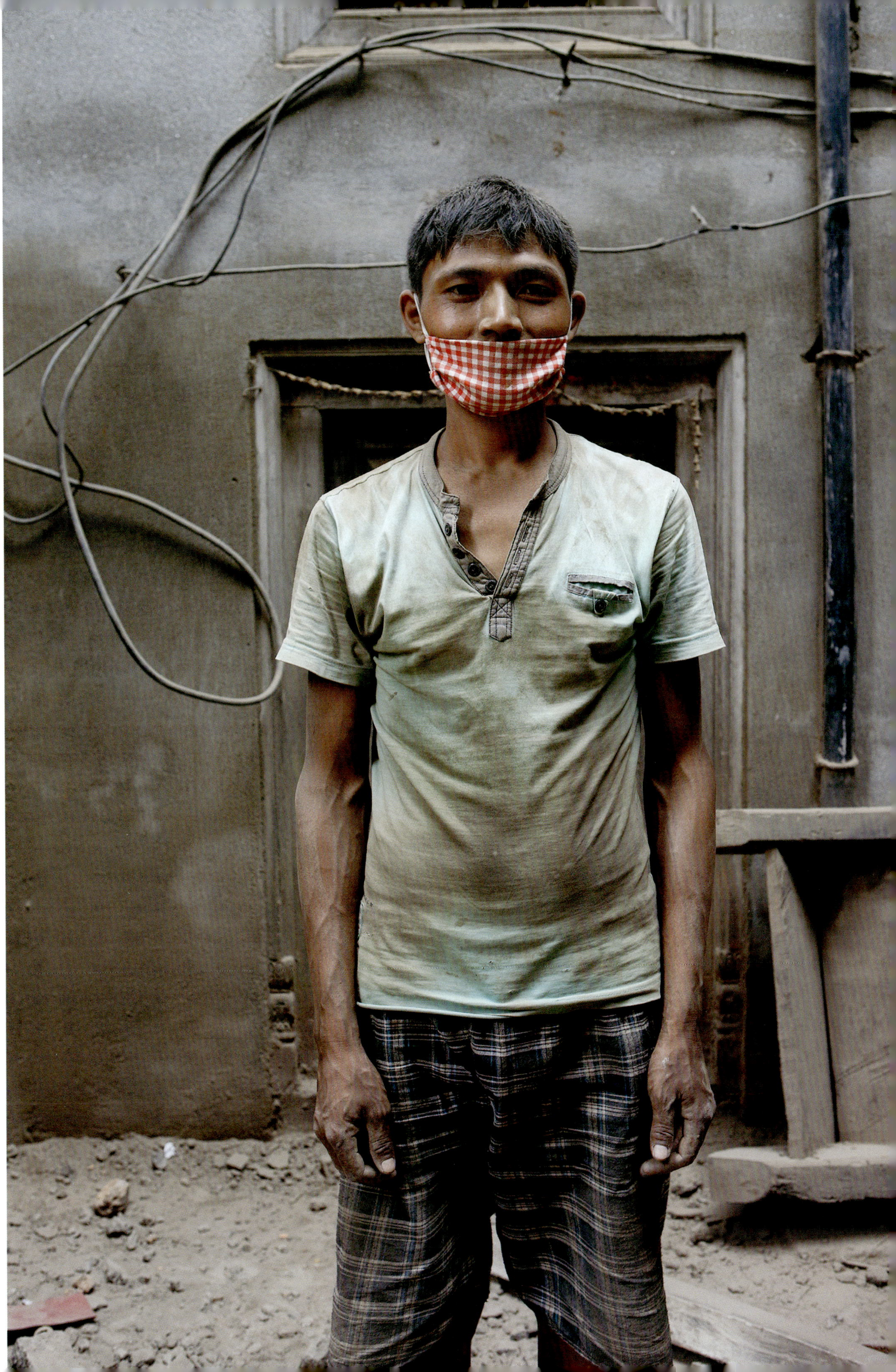

Ramesh Lama,
30,
Ramechhap
district.

Jit Bahadur shovels debris from the third storey of a house partially demolished in Bhakatpur. Assisting him by pulling the rope tied to the lower end of the shovel handle is his neighbour, Prakash. While it is the monsoon and humidity fills the air, freshly dismantled brick walls create clouds of clay dust.

Bhaktapur
resident,
Sagar Thapa, 15.

Mourning and resilience

— Pramod Mishra

Amidst the COVID-19 pandemic, I'm trying to remember another disaster, Nepal's 2015 earthquake. Both seem to be nature's wrath on human excesses, human carelessness, and human pride in their all-conquering ambition. In the noise of the epidemic, the Nepal earthquake of 2015 seems a distant thunder.

But it remains vivid in my memory because when it occurred I was asleep in Chicago, gone to bed observing a one-meal fast in memory of my mother who had passed away eight years earlier on 25 April. I woke up to mourn 9,000. The picture unfolding the next day like the photos in this collection overwhelmed me with grief. I wrote the following:

> April 25 since 2007 has been for me a personal day of mourning for my stateless mother, and this year on the day will also be part of annual mourning for thousands of 'stateless' Nepali men and women who perished in the earthquake. Every year on this day, I remember my mother through dietary restrictions, reflection and sometimes visit to a local temple in the Greater Chicago area. I recall my mother's stateless life, her indomitable will in defiance against her own family and caste and my father's family and caste to rebuild her life among strangers who became her own. She lived her life the way she decided to live despite adversities brought about by caste, clan and states. Like other years, this year also, I had stayed late in my study until midnight on April 24, thinking of her, going to bed with thoughts of the next day of mourning and visit to the temple. I woke up next morning to mourn thousands.

I know that survivors, with their indomitable will, grit, and resilience will rebuild their lives and houses with the help of their neighbours, country folks and the international community even though the deceased will never return, and their loss will never be repaired. But this is also an occasion to reflect on their special kind of statelessness occasioned by being for centuries under a self-centred ruling class of one political provenance or another who have done everything fair and foul to grab the power of the state to rule them by using the Nepali verb *hakne* (drive), as in *desh hakne* or *party hakne*, rather than serving them by working for their welfare.

Earthquake experts very often assert the cliché that earthquakes don't kill but buildings and structures do. And it was the duty of the state and the government that occupied the state to prepare people to act before, during and after a temblor in a quake-prone zone. The Nepali governments, whether run by Ranas, Shahs or Bahun-Chhetris, family oligarchy, family dynasty, or multiparty political hill-caste system, failed miserably to prepare the country and its people to face what was an inevitable and predictable natural disaster. How is it that they couldn't hear the noise the cohorts of national and international seismologists, environmentalists, earthquake experts and journalists had been making in the media and conference halls for decades?

Kathmandu's architectural landscape and town planning represented the chaos of Nepal's political landscape. How did Kathmandu become a slum-city over the decades? Even if we allow that the Malla-era architecture and town planning of old Kathmandu, old Patan and old Bhaktapur could be forgiven because centuries ago the Malla rulers had no idea about the future nor did they have the technology to think ahead, how can one forgive the modern ambitions of the Ranas, the Shahs and the three-decade rule by multiparty mandarins? Didn't the Rana rulers build massive palaces in imitation of the best of national and international architecture? Didn't Mahendra and Birendra rule with an iron fist in the name of modernisation? Haven't the multiparty leaders ruled in the name of democracy? Then, why allow Kathmandu to grow into a slum-city with nary a thought about implementing civilised town planning, quake-resistant architecture and disaster-management and preparation?

The one answer that comes immediately is to focus on the rulers themselves. There has been a representative government since 1990 but there hasn't been a representative government in the real sense of the term. Each leader, each party, each ruling ethnicity, like their predecessors, is consumed by the question how much it will gain or lose if political changes occur.

How many political leaders died in that Saturday's earthquake? How many high-level bankers, bureaucrats, security officials perished in the quake? Even though one can't and shouldn't generalise about the class orientation of the victims of the quake, one can safely say that, save for exceptions, it's been mostly common folk who lost their lives. When my friends from far and near ask, 'Are your family and friends alright?' I say, yes, they are fine as far as I know. Then, I add, 'But that is hardly a consolation.'

Take, for example, the need to contribute to the relief effort. There was scepticism in the air among Nepalis whether the contributions, especially money, would reach the needy on the ground. Many openly said that the disaster had become a god's gift to the middlemen, officials and politicians in Nepal to skim and grow even richer from the worldwide earthquake relief funds. Many wanted to help the victims directly rather than go through intermediary agencies. Why was that? Why was it that Nepal and its ruling class during their years, decades and centuries of rule allowed moral degradation, creating a political climate where those in power and positions had only one concern—how to fill their own personal, family and clan coffers? Would the reputation of Nepali ruling class further deteriorate or rise in this generational disaster?

The natural disaster was terrible for the people who have been directly hit by it. But it was also a challenge and an opportunity for all Nepalis no matter where they live to think about the political climate of caste, clan, ethnic hierarchy that has made us selfish and callous. The best instincts of compassion, generosity, common humanity and fellow-feeling the earthquake has brought to the fore ought to warn us against baser instincts not only in ourselves but in others as well. It should teach us to think collectively for short-term action and long-term vision for the greater common good.

What ensued after the earthquake? As in this pandemic, the political class behaved as it did during the disaster: fishing in troubled waters, both politically and economically. Mismanagement and corruption stood out as political culture compounded by the compromise among three hill 'high-caste'-dominated political parties that brought forth the flawed constitution on a fast track.

I am not the fund-raising kind. I prefer to give away my own meagre money rather than ask others to donate because of three reasons. One, most often, I have not found a worthy enough cause to ask others to empty their wallet. Second, even when some cause presented itself, I became sceptical of the medium through which donations could reach the deserving, especially in the age of more mediums than perhaps causes. The mediums themselves have become a replica of government service, a major source of employment, salary and perks for the ambitious young men and women, skimming off much of the funds. But, most importantly, I am always involved in my own affairs with little time to spare for others. But that was not so when growing up in a Morang village and that was not so with the 2015 earthquake. Overcome with grief and guilt at my own privilege and safe distance, I set up a table at the local Hindu temple in Lemont on the occasion of a South Indian cultural programme. I collected about a thousand dollars and gave it to Bishnu Phuyal, a local organiser. And I wrote more about it.

The 25 April earthquake constituted the tripartite of destructions that have wrecked Nepal's mythopoetic in the 21st century. Needless to say, all three have been traumatic and extremely painful, especially for those caught in them. The 1 June 2001 royal massacre came as destruction from the top of Nepali society which had been hollowed out in its core by its many contradictions. Centuries of instilled and cultivated faith in the system symbolised in the royals lay shattered in its aftermath. The Maoist insurgency lasted 10 years and didn't come as suddenly as the earthquake or the royal massacre but it was a destruction of the rot launched from below. People oppressed for centuries found voice in the insurgency. The earthquake came as suddenly as the royal massacre, but this natural calamity destroyed entire hill villages and most of those who lost lives in this quake were common folk.

While the quake may have been non-ideological, its effect was very much political for it exposed the inefficiency and incompetence of the civilian government. Stunned, Nepalis responded to the quake the way traumatised do—emotionally.

The quake also provoked nationalist sentiments, some ennobling while others were conceited and knee-jerk. Most voices in the media invoked Nepali pride and self-respect, bending not breaking, fallen but not defeated, etcetera, etcetera. Even otherwise, Nepali-speaking folks of a certain class and sentiments wear their nationalism on their sleeves. At a time they found themselves without the wherewithal to cope with a disaster of such a magnitude and saw outsiders, especially India, swiftly sending planeloads of help, their helplessness surfaced as an emotional response of national pride and resentment at their own inability to cope. And when they saw and see the civilian government criticised, they reacted as if the government was their uncle. But this attitude won't help. A nation needs to learn gratefulness for support from outside.

Even those who hadn't heard much about Nepal put a finger on the map and expressed words of sympathy. In this sense, the earthquake may have made the old state further vulnerable, badly damaged as it had become after the royal massacre and the Maoist insurgency. But Nepal's new sovereignty emerged ever more strongly because the earthquake forced the country onto the consciousness of people the world over. This goodwill of the world community could have been overwhelming at times, but it would not have been wise of Nepal's political and nationalist class to squander it by resorting to cheap nationalism.

No, ultimately, Nepal's political class didn't squander international goodwill. It used it to forge the 16-point unity of 'upper-caste' leaders of ideologically divided parties and brought forth the flawed constitution over the deaths of more than 50 Madhesis killed at the hands of the security forces.

Disasters and pandemics are as old as the earth but those who rule haven't learnt lessons to work together for the collective good. Yet, those who find themselves caught in them eventually pick up the pieces and move on. The people rebuilt, with help from others. Humanity came together; but it also remained apart.

स्कोप

PATAN

Overleaf: In the Chyasal neighbourhood of Patan, Buddhist stone shrines called *chaitya* and prayer wheels were damaged by falling brick and debris from the surrounding houses. This section of old Patan was damaged more severely than other neighbourhoods.

The famous Krishna Temple in Patan Durbar Square.

The remains of the Char Narayan Temple. Also called Jagat Narayan, this is perhaps the oldest temple in Patan's Darbar Square, and unlike others there, is constructed primarily of brick.

Overleaf: Temple roof struts salvaged from the Char Narayan and other temples in temporary storage in the Patan Museum.

Spread on pp. 130–131: Binod Byanjankar prepares to rebuild his restaurant, Byanjankar Khaja Ghar in Chyasal, Patan.

MINSK
MINSK

Unmoored

— Pranaya SJB Rana

When I was a child, I believed in the infallibility of things. Parents were always right and teachers never wrong; dogs were friendly and strangers kind; love was simple and friendships strong. Each day was a new beginning, shorn from yesterday, cleaved from tomorrow.

Everything was solid, stable, like the ground beneath our feet.

And, just like that, everything fell apart.

When the ground surges up to meet your feet, there is nowhere to run. The earth roils like an angry sea and you realise you are nothing but a boat, bobbing along on the surface, content that all is well in the deep.

We ran, of course. How could we not? Millenia of evolution has wired into us the basest of responses to any unexpected occurrence—fight or flight. There's no fighting an angry earth. All you can do is huddle in a corner, arms over your head, praying to a god you aren't sure exists.

Or, you run.

We ran out of our homes, stumbling along the way. We ran onto the streets where the pavement cracked and the electricity poles swayed to and fro like reeds in the wind. We ran onto fields and courtyards, holding onto each other, friends, family, strangers, anyone, seeking one point of stability in an increasingly unstable world.

Over the coming weeks, we would run often. Each time there was a rumbling deep and low from beneath the earth, followed by a jolt that sent every heart palpitating. While sleeping, bathing, eating, we ran. Even when it rained, we ran.

If you couldn't trust the ground to stay moored, was there anything left to believe in?

On the very first day, I remember a house that had collapsed, trapping three or four people underneath. It had flattened as if a giant hand had pressed down, strong and hard. It didn't take long for people to form a human chain, digging up debris. Those who emerged from the ruins were born anew. A broken bone, a few scrapes, they lived to narrate their experiences of what it feels like to be buried alive. The sky goes dark, the air recedes, and you go gently into that good night.

An army man showed me his palms, raw from having leapt off a balcony two floors up. He had landed on his fours, regressing into the animal that made us all who we are today. And, here he was, pulling dirt, bricks, rocks and stones with his palms bloody.

Two years later, another man would show me his palms again, raw from lashing down the tin roof to his one-room shelter that kept blowing off in the monsoon rains. When the sun was out, it was unbearably hot and at night, it was bitterly cold, he said, laying his palms out flat.

He'd had a two-storey house before, just a few kilometres from Chautara, mud and brick, with a thatched roof shed for his four buffaloes. When the house fell, it killed his mother, it killed his wife, and it killed his four buffaloes.

Two years later, he was still living in a tin house that got unbearably hot during the day and bitterly cold at night.

We were lucky that our houses stood. We were privileged. We lived in Kathmandu. We had money. We had neighbours. When everything else was falling apart, our centres held.

And, yet, lying under a strip of tarpaulin as the rain pattered above, my dog howling into the wet night, there were thoughts that cannot be spoken, lest they be realised. Suffice it to say, mere anarchy was not loosed upon the world. But every night, we rested our heads on hard pillows with the possibility that there might not be a tomorrow to wake up to.

Years later, when woken in the middle of a European winter night by the rumbling of a passing truck, it would strike me, like a jolt from the earth, that faith once broken stays shattered a long time. I knew then that we were all marked, imperceptibly perhaps but marked nonetheless, by the trauma of a shaking earth.

There are many of us who carry that weight, and we will carry it a long time. There's no telling when that long-suppressed experience will resurface, reshaping itself like all memory does, becoming something instantaneous. So that whenever an airplane flies particularly low and shakes the foundations of your home, you are halfway out the door, your limbic system taking hold, consciousness be damned. You only realise, after the fact, that you have been conditioned, through weeks of aftershocks and new tremors, that the earth can be as fluid as the Narayani river.

We moved on, like humans are wont to do. We forget pain quickly, which is what allows us to get on with the daily drudgery that we call living. Our loves evaporate in a plume of smoke from a funeral pyre on the banks of a slow-moving river. We pour ghee on their smouldering corpses and watch the fire blaze. Then we go home and we sleep, only to be woken occasionally by a jolt to the heart.

They told us there was a lesson here somewhere, in the rubble of homes and the detritus of hearts. We were supposed to come out of it better, more empathetic, less fatalistic. I am not sure we learnt anything. I'm not sure there was even a lesson there. Perhaps we've become inured to death but that was only natural when death happens only to other people.

Thousands die every year when buses careen off highways into fast-moving rivers. Thousands die from having too little to eat, too much to eat, too little to love, too much to hate. Thousands have died in the last few months alone. Thousands die every single minute, all around the world. For some, death is romantic, an interregnum from the banality of life. For others, death is a spectre, looming over your shoulder, there all the time. For them, death is as real as hunger.

I was seven years old when my father died. Over the years, family members would continue to die with alarming frequency. But during the earthquake, no one I knew perished. My home didn't collapse. I wasn't pushed into poverty. The disaster was something that happened to other people. We felt the heat, but there were others who burnt.

And that's how it has always been in Nepal. The war happened to other people. The accidents. The floods. The landslides. The riots. The dysentery. The malaria. The coronavirus. It all happened to other people. We went out and took pictures. We went out and wrote newspaper reports. We spoke to them and even though we looked them in the eye, we always quickly looked away, ashamed. We always knew that our pain was vicarious.

Those of us whose belief in the foundations of our faith was shaken were those who were lucky enough to be able to trust in things. For those whose homes have been swept away by a landslide or a flood, whose fathers and brothers disappeared in a war that belonged to someone else, whose daughters never came home from Kathmandu because they'd fallen into the Trishuli—what is there to believe in except for the capriciousness of everything.

When the earth shakes and we stumble over loss, we turn our heads to the skies and howl—why me? Everyone else already has the answer—why not?

KHOKANA

Samyam Dongol, 3, and his friend Shining Dongol, 2, are settling into enjoying Shining's family's new dwelling. The bed frame had just been brought in and a few more pieces of furniture rescued from their uninhabitable old house were arriving. Then it will be time to work on Samyan's family home.

Budhi Gandaki

— Avishek Shrestha

In 1998, the football world cup in France was already a week old. Yet the *GorkhaPatra* national daily my father was reading wasn't done discussing it. From my view in our living room, you could only see the cotton clothes he was wearing on this warm summer morning in the western foothills of Nepal. The rest of him quietly buried behind the cover pages of *GorkhaPatra*, where the news reporters were loudly playing out the world cup.

A hint of cardamom, or ginger, I couldn't tell, swept through the living room otherwise filled with an air of mild anticipation. In the kitchen, my mother was stirring sugar in the milk tea she'd prepared for us. Its warm scent as if carried to us by the sound of every spoon clink. Our tea was still swirling around the glass rim when she brought it over. It had to be card-a-mom. My mom's favourite card, I suppose.

An unmistakable atmosphere of a slow, lazy Saturday was fast developing. Besides the occasional shuffle from the kitchen, our water buffalo had already moo-ed twice from the backyard, demanding breakfast. A motivated clan of neighbourhood roosters, in the meantime, had taken turns to remind the not-so-motivated humans to wake up. Street dogs and domestic goats, then birds in the nearby forest also chiming in with their views on the matter.

Otherwise, the only soul you could hear was the ever-present low hum of the BudhiGandaki river. Our home along with most of the neighbourhood sat on a gentle cliff next to it. The cool morning breeze from the river periodically washed the stone-paved street outside, and at moments fluttered *GorkhaPatra*'s sunlit pages. Graciously making its way from up north, every sip of our tea that morning was inching BudhiGandaki closer to meeting the Ganga river in India. As far as I could tell, it was in no real rush.

You could've surely sensed a greater urgency if it wasn't a Saturday. By now, kids like me would've been dressed up in sky-blue shirts ready to go to the local school. All public school students throughout the country wear the same colour even today. At that hour, the street outside our home would've seen waves after waves of the emblematic sky blue. Sometimes paired with clean combed hair and black polished shoes but oftentimes with just flipflops on a ragged look.

No matter your looks, that bit was an incredibly enticing part of attending school—marching across town as part of the blue wave with a sense of belonging and purpose. By the time you reached the classroom, however, the lunch break couldn't arrive sooner. All the kids would spill onto the streets as soon as the school bell rang for recess. And some would decide not to return for the afternoon classes altogether. The best excuse that worked for a summer day—'*I have to help my parents in the fields before the monsoon rain.*'

For a Saturday though, my sole intention past the morning tea was to convince my father to go to BudhiGandaki for a swim. The green wave of the river was very enticing. Even if its currents were annoyingly strong, and you often feared there was no returning if you went too far; those were still easy compromises for an ice-chill blanket of BudhiGandaki before the sweat of noon. The best excuse that worked for a summer day—'*We have to beat the heat before the river swells from the monsoon rain.*'

So, there I was, waiting for an opportune moment to formalise my proposal. My father was finally shuffling through to the last page of *GorkhaPatra* when—'*tmnt, tmnt, tmnt, tmnt, tmnn...*' the peculiar set of five notes called attention from the radio nearby. '*Yo Radio Nepal ho.*' This is Radio Nepal, a voice announced. Naturally then, my father peeked past his *GorkhaPatra* and signalled me to increase the volume. Seemingly attending the newspaper, he was now simultaneously reading the radio broadcasting live from Kathmandu. Surely now, BudhiGandaki wouldn't be a priority.

Back in the late 90s, the nearest motor road from our home was a day's hike. Three, if you'd rather enjoy it. Then another day of a bumpy bus ride away was Kathmandu, where the *GorkhaPatra* press and the Radio Nepal station were located. The radios in our town captured only one station, which offered the latest world report. While the paper—a few copies hand carried weekly from Kathmandu by the local postman—supplemented the news with further details and colourful pictures. Therefore, as far as my father was concerned that morning, the world cup was played on schedule and reported timely: not a week early, not a week late.

At the turn of the century, it looked like nothing would ever change in our part of the world. Neither the all-knowing Radio Nepal nor the all-showing *GorkhaPatra* had predicted any evolution. In less than a decade, however, there was a non-stop bus service to Kathmandu from our town. And you could reach Qatar the same day you touched BudhiGandaki. The streets had more smartphone shops than radios. And a variety of newsstands sat where once a lone photo studio hung a sign that claimed in bold Nepali—'*Colour portrait ready in just ten days!*'

But we weren't quite there yet that morning. In our living room, *GorkhaPatra* was still playing the world cup till the final whistle. Radio Nepal was still narrating the stories of Kathmandu, its king, his ministers, and their foreign diplomat friends. And BudhiGandaki was still only a stone's throw away, as noon approached tantalisingly close. Without us knowing, the blue

sky outside had unexpectedly given way to a group of dark, murmuring clouds. Only a few folds of hills up north from our home, they were covering the peaks of the Ganesh Himal, then the hilltops of Maanbu and Laapu, then Laprak—where the earth would eventually give way on a Saturday similar to this. It was too early in the year for the monsoon rains. But any rain up north bore heavily on the river down the cliff.

Then my grandmother, or Aji, as we grandkids call her, walked into the living room to break the news—'*Babu, run outside and tell your friends no swimming today.*' She was now as old as BudhiGandaki, as slow as its pace, and as wise as its soul. While my restless mind had already raced across the town. Past the file of mules destined for the northern valleys, past the band of fabric stores and rows of food stalls, past the goldsmith shops, tailors, and farmers, market, past the Hindu temple and the Buddhist one, and past the pyramids after pyramids of grains and powders of spices neatly on show on both sides of the street. At the edge of the town, just far enough away from the bustle of life, there was an open, green field by the river. We could still play football there even in the rain. Or simply lie by old BudhiGandaki and watch it continue to hum in its ever presence.

BUNGAMATI

Facing page: She and her husband were too shy to let me know their names. They lost pretty much everything except the footprint of where their house stood. They have cleared away most of the debris and need to make a safe and secure temporary shelter as they prepare for the birth of their first child. While they have so little, they have each other, their health and the birth to look forward to.

Babu Kaji Sarki stands on the ruins of his house, the house where he was born 65 years earlier. He, his wife, their son and four grandchildren lived here. All of their belongings and food supplies were buried in the rubble.

Kiliz Tuladhar, 18, wood carver/sculptor, is one of 30 or 40 young men and women of Bungamati who volunteer from 6 to 11 in the morning every day removing debris from the neighbourhoods. No one is paid, but at 11 am everyone receives the morning meal cooked collectively by the town.

Jiten Khadka, 21, and Raman Bajracharya, 18.

Reg : 16899
Give children, women the opportunity to get
You can involve in BFN programmes many
Make a donation .... even the smallest amount
Research about community, culture, religions.
Volunteering, share talent, skills, knowledge.
Home stay.
If you are interested in our organisation and
B.B.S.
B. Ed.
B.A.
P+V 25

Riding the rubble

— Rabi Thapa

Jyoti

She hasn't been able to sleep since. Which is a problem, because he hasn't been able to, either. The other night, she sat bolt upright in bed, nerves jangling like her whole body was a funny bone, all elbows and edged teeth. She scrabbled for the light behind her, slipping in a sleeve of sweat, only to be confronted with his pockmarked, quizzical annoyance.

'Ke bho pheri...'

'Was that a -'

'Uff, for god's sake chill out. Are you mad? I *coughed*.'

And he turned his back on her. For a few seconds she stared at him, appalled by his callousness. He had been fast asleep after a long day's work, but still... Then the bed began rocking again, there was no question this time. But even as she reached out for him she could see his body shaking just that bit extra, thrusting back and forth like a jack in the box. She could see his teeth in the dark, then his muffled hoots of laughter. She grinned, *hyaa moolah!*, and pulled him close.

Rishav

The tremors made him nervous, too, but they were good for business. People kept coming to him, in waves as predictable as the tremors were not, to ask for supplies: tents and water principally. And with his friend in the Ministry of Supplies, selling off Chinese tents that had been hoarded from the last disaster was easy enough. Even so-called refugees from the villages, when pushed, came up with the money. And water? All people needed was the packaging, it didn't matter what was in the jars as long as you could see through it. They needed reassurance in these times, they needed to feel that they were working through their *situation*, and he was the man for any situation.

Sharada

'Mummy, there's a worm in my glass!'

Her eight-year-old was presenting the offending item to her, demanding an answer. But her enthusiasm suggested morbid curiosity rather than revulsion.

'Well, the water came from the river, and many things live in the river, because they need to drink too, see? So, you're sharing the glass of water.'

And she made a worming motion, gulping down the air around her until the girl burst into giggles and started imitating her.

As she took the glass from her daughter with a half-concealed grimace, she decided she would save the worm so she could try and get a discount off the next jar. In the meantime, she needed to get to the district administration office to make sure their names were on the list of the affected. People complained that the system didn't work and that they weren't getting what was due to them, but wasn't it better than nothing at all?

Durga

He knew the widow lived by herself, and spent whole days moving bricks from one side of her yard to another. He'd seen her. Still, she should have made the time to come and see him about relief supplies. Everyone had to look out for themselves in these difficult times. Take the hotel down the road. Every time a relief truck roared over the crest of the highway and turned into the yard before the district office you could be sure either the husband or the wife or one of their boys would be waiting with an aggrieved expression on their face. They wouldn't leave until they had added to the stack of sacks they displayed, almost proudly, in the stairwell of the 20-room hotel they had built with the husband's Saudi savings. He had no doubt he was paying for relief rice drawn from the supplies he'd signed off for when he headed there for dinner.

It was his job to make sure the government provided for all, of course, which was why he would look out for the widow as well. She could make do with one of the smaller packs that had come in, and the hotel could use some lentils; hadn't they complained they were out? Dal cost twice as much as rice, but that was the way it rolled. He'd even leave aside a sack for the trader who arranged extra fuel for the office, even though he'd moved away a decade ago. He was on the list, after all.

Daniel

They were feeding the five thousand. Praise be, they were feeding the five thousand! It sounded dramatic (it sounded Biblical, everything in the Book sounded Biblical, wasn't that funny?), but it was the literal truth. He was sad to see the poor Nepalis suffer so, but God had provided, through the good Christian brothers and sisters of Cedar Rapids, Iowa. Their gratitude showed on their sweaty faces as they pushed to the front—well, it wasn't always clear that they were grateful, but one had to understand these were desperate people in desperate times. At least their leaders had the sense to express thanks in the correct way. Banner held up for the photo, speeches from both sides, smiles all around, what a perfect day! He was certain the good old boys up in Des Moines would feel the same glow in their hearts he was feeling right now, once the dispatch went out and the numbers went up on the site.

Stop pushing, goddammit!

It was hard, sometimes. But it was the Lord's work, and if he wasn't doing it, never mind what these poor people would be doing, what would he be doing back in Mechanicsville, Iowa?

Prasit

It was very distracting, the thoughts that kept pushing into his head as he wrestled his way through the crowds to fit the Christians, their enormous banner and enough of the locals into the frame. Mock captions formed as he clicked the camera—*Rice bowls for souls! Quakers for Quakes!*—undermining his intent and distorting his vision.

He took one more and composed a few words to sum up the lot.

Iowa church presents much-needed supplies to earthquake-affected community.

There. That would do fine. And wasn't it as much the truth as anything else? He would keep his doubts to himself, and submit this, along with a few portraits of the mountain community sweating it out in tents in the monastery grounds. It didn't matter where the aid came from, as long as it came. It wasn't as though every other religious community wasn't marketing itself in exactly the same way.

He was a photographer, and it was his job to get the word out. Not the Word, he reflected, as he allowed himself to be led by the sleeve by one of the community leaders, who was insisting on solo portraits for a select few over the sacks of flour that had come in. He was shooting blind by this point, but it didn't matter. His by-line in the Delhi Times was secure.

And if he hadn't gotten back to the American correspondent in time, so what? She would have been travelling with her own photographer, and he would've just been a glorified, overpaid peon. Yes, he might've been able to buy that wide-angle lens he'd coveted for so long but, no, he would scrape his way to it, by-line by by-line.

Nayan

There was only a moment of hesitation, easily put down to the difficulty of transmuting the sing-song syllables of his mother tongue into the internationally intelligible. In that split second he made the decision to lie to the foreign journalist and continue to do so whenever he encountered an inconvenient truth. For, if there was comfortable, familiar outrage in telling her how angry these people were at the government, what use was it for her to know that a portion of the rage belonged to her? So, they wanted to know how much money the foreigners were making off this gig, of course they did. He tried to mollify the furious old man with dirt driven into his wrinkles by telling him that the coverage would bring in the money that the *sarkar* had failed to hand out, and smiled back at the journalist, telling her he was trying to console him. She bought it, hook, line and sinker. God forbid that she cottoned on and felt shamed enough to terminate her assignment and his.

He wasn't a fixer, he was a diplomat! And so, a fixer! And not, what had that wretch called him at the end, a bloodsucker, living off the suffering of his own people. A bloodsucker! Everyone had to live, and everyone did what they had to do. It was true that he was making ten times as much as he might have done if not for this cursed earthquake. But it was also true that his work would help flood the coffers of his useless government. For the sake of this old man, he hoped some of the money would clear the rubble that still buried his family, a week on. But that was a story for another day. This one would have his name on it.

Jo

It was the one photo that stayed with her, after the dust had settled, after all the stories she'd read about this faraway place she'd always wanted to—still wanted to—visit. Children, walking past the ruins of buildings that were once the homes of family, friends, neighbours, walking on the still-quivering substrate of a land that had shrugged off, so disdainfully and unmindfully, the dreams and aspirations of millions. Walking to *school*. The accompanying narrative didn't put it in so many words, perhaps it hadn't even occurred to the writer, but these were traumatised kids, going to school a fortnight after a 7.2 Richter earthquake that had killed 10,000, to sit in a classroom, perhaps with a hole the size of a Mack truck in one of its walls, and learn Maths, English and Science. Maybe even the science of earthquakes. To prepare for a future that they, or at least their parents, still believed in. After all this. Never mind what had just happened. Life went on. Life goes on.

CHAUTARA

CASE

SEEKERS

Rama Tamang (in pink sari), 35, working with her mother, Urmila, and Man Bahadur Tamang, 36.

Filming Langtang: In the aftermath of the earthquake

— Kesang Tseten

In the immediate aftermath of the 2015 earthquake, as the tremors continued to jolt us, physically and emotionally, I got calls from two foreign television teams offering me work.

I was staying home, with my son whose boarding school had closed, while my wife was at work, anxious and reeling, like most of us in Kathmandu. The experience of the ground quivering under one's feet jolts the sensorial experience of life as few things do.

I was also feeling somewhat at a loss as I could not join up in the numerous relief efforts underway by civil society, aside from making modest financial contributions here and there. Thus, I declined the offers of work from the tv units covering the quake.

I didn't want to make a film about the earthquake either because I didn't want to make a 'disaster' film. Most times, when I decide to make a film, I do not do it believing my filmmaking or the film that emerges will help the subject of my film. I do it because it is what I do, my raison d'être: to encounter a subject, immerse myself and inhabit the world of the film's subject, the film being the result or the ordering of that encounter.

The film might even be of benefit to the subject or society in some way, but rarely, and in any case it isn't *primarily* why I do it.

The reason I did make a film about Langtang was not because it suffered the most or that it was hit in a unique way. The collapse of a glacier from Langtang Lirung, ironically a local protector deity, decimated more than a quarter of the 700-plus members of the community, burying Langtang village under 25 metres of debris.

I was thinking about the one Langtang-ba I knew when he called: Gyalpo, the cheesemaker in Kyangjin, the first cheese factory (1954) of a dozen that Swiss development aid had established in Nepal. He had, like the entire community, been evacuated to Kathmandu. He wanted some goat's milk for his sister, who had suffered a serious injury to her pelvis. She was taking *ayurvedic* medicine, which the *vaidya* recommended be taken with goat's milk. A true cheesemaker, Gyalpo remembered seeing goats on the slopes of Budhanilkantha, where I had been living for two years.

I dispatched someone more capable than me to find the goat's milk, which I took over to the Yellow Gompa near Swayambhu, the site of 116 tents, one for each Langtang family evacuee. Some days later, I took over another litre of goat's milk, which seemed to have helped, as Gyalpo's bedridden sister told me in a whisper.

Curiosity got the better of me. Just *how* would I make a film about the Langtang community? On an impulse, I asked a young filmmaker, Nima Lama Gelbu, who had been interning with me, to begin filming at the Yellow Gompa. Soon after, I asked someone on the Langtang committee for permission to film. He instantly approved. I was trying to decide whether to continue filming or not when the same person changed his mind. He said he had heard I made 'international' films. He said we should share the profit from the film I hadn't yet decided on making at that point.

It is true my films are often screened internationally; untrue that I make a lot of money from that or from making films. I consider myself lucky if I get an adequate grant that pays for the cost of making the film, including a modest remuneration. In this instance, until then, I had not applied for funding from anyone. Subsequently, I was fortunate to receive grants to cover most of the costs.

Thus, my initial response to his suggestion was part irritation, and part indignation. Here I was, trying to do my part, to do something right. But then I thought: what was it to anyone that I was going to make a film? Should they thank me, embrace or salute me?

It was my problem. The truth was, I was in it for my filmmaking self. It could be useful, it could help Langtang. In any case, it was not my intention to make an advocacy tool, to raise money. The person proposing the 50-50 share on the profit didn't know what he was talking about, so it was pointless recriminating against him. So, I continued filming. To not do so seemed irrational, to be taking it unnecessarily personally.

It also occurred to me that the Langtang folk, with homes and thriving businesses in Kathmandu and Langtang, were not the usual poor villager we generally make films about. They were worldly, savvy people, and understandably sceptical about a filmmaker's project, and why not.

Retaining my ambivalence not to make a 'disaster' film, I set myself the challenge of doing it differently. I decided I would stop if the Langtang community asked me to do so. I made a modest contribution to the relief fund, without announcing it, but the word probably got out and that might explain why no one asked me to stop.

No one disapproved when my colleague and I trekked up to Langtang and Kyangjin, even as the earth shook with after-tremors, following the Langtang men as they searched for the bodies of their loved ones as well as recouped what they could of the valuables left behind in their homes buried under the debris. We filmed their daily activities as refugees, their internal arguments in the event the government gave them land to apportion amongst themselves, understandably fractious. We accompanied them on their visit to the government dairy corporation to plead for write-offs or leniency in paying back money advanced to the yak herders against delivery of milk for cheese production. And when they asked for rebuilding of the cheese factory, given the total washout in tourism income, which 95 per cent of the community depended on.

We filmed an entire year, in the Langtang Valley, as well as in Kathmandu, until the first yearly commemoration of the quake. By then, they had already rebuilt most of their homes and resettled to life back in the mountains. For the first time, a community member expressed appreciation to us for staying with their story long after the topicality of the catastrophe.

As for the film, the task was to discover the story I wanted to tell. I realised mine was to show who they were, the earthquake serving as a pretext.

Not entirely though. We had filmed adopting a mostly observational approach, which was to eschew direct interviews, except in the case of Gyalpo. Since he was our entry to the story, with whom we travelled during the filming, it would have been odd, an artifice, if he did not speak to the camera. I chose not to interview other folk, however, as what could they say about the devastation and personal losses—some had lost entire families.

Nonetheless, an awkwardness set in about not engaging with them in some way. That triggered the idea of setting up a make-shift studio at the Yellow Gompa, inviting a member from each family tent to speak to us, saying whatever they felt like saying. Unsurprisingly, the 85 people that took up our invitation spoke about the disaster one way or another.

Excerpts

An elderly man:

All the youngsters died,
only this old man survived.

Two sons, two grandsons ...I'm all alone...I should have died.

A young woman:

I was brought here, evacuated by helicopter...

There's no one left. Just me.

A man who lost his wife:

I lost my wife. That's all.

As cheesemaker Gyalpo says, at the beginning of the film:

From ancient times, our Langtang was known as a sacred valley...a heavenly realm.
... tourism thrived, and we all prospered.

All was going well when...whatever the mountains gave us, the mountains snatched away from us.

While editing the film, I did not initially use these interviews as they had been done on a lark. But on viewing our rough cut, the film had seemed flat. It was one thing not to dwell on peoples' suffering, as in a 'disaster' film, another not to make any reference to what had happened, I reflected. So, a selection of the interviews found their way into the film, interspersed, as potent and direct emotionalities.

At the first screening of the film, only for the community, the Langtang folk expressed their thanks and their approval. The telling comment, however, was: how about making part 2: the story of how Langtang has risen.

As a filmmaker, I had been set on making a film that was truthful, not for the purposes of advocacy, such as a call for help. I didn't quite understand this disinclination to remain in grief, I thought with some disappointment. How could they be so eager for tourism this soon? But that was me (just as not wanting to make a 'disaster' film was me). It was presumptuous on my part to think so, I realised. What can anybody *do* or *be* that is commensurate with the scale of grief and loss that had befallen them? Perhaps recognising this, they believed rebuilding their livelihoods, insofar as it represents sustenance, allows the possibility of helping themselves materially and more.

CHARIKOT BAZAAR

Nursing students take a break from their studies to say hello from their classroom. Like so many youth across Nepal they are energetic and optimistic about the future. This nursing school in Charikot Bazaar is sandwiched between buildings in various stages of collapse.

Anisha Shrestha, 7, studies in Grade 2. She and her brother, Ganesh, walk through the ruins in Charikot twice a day while going to and coming back from school.

The refuge of tremors

— Prateebha Tuladhar

Are you alright?
Can you file?
-D
As the rattling drummed down temporarily, I checked my phone to find this note from my bureau chief in my email.

It had taken me a while to get to my phone because when the ground beneath us started shuddering, my dog and I were on the veranda, and my first impulse was our safety. I had tried to make us duck under the wooden bench there, but how would a dog understand that? It was not before I managed to get us to the safety of an open space that I could go back and grab my phone on my desk.

Going back into the house meant navigating the debris of a broken mirror in the corridor, souvenirs, mementos, china vases, and other paraphernalia out of which life spills. Stowed-away pillows and blankets from the shelf just above my desk had tumbled down, forming a mountain and closing the cover of my laptop—it hadn't broken.

My computer and my phone. Those had to be rescued first.

I had to file! The entire neighbourhood had gathered in what was the only open space around, belonging to Chejum Auntie, a neighbour who grew maize and vegetables. We trampled her vegetable garden, women and children huddling together, men standing nearby, as the ground continued to shake.

We started calling up people only as much as connections allowed us—siblings, coworkers, sources. Luckily, emails were working. And that's how the story broke: 'Nepal shaken by a powerful earthquake. Details on the magnitude of the quake and extent of damage yet to be confirmed.'

With electricity cut off, batteries were draining, threatening to sever communication. Some of us made quick trips back into our homes to retrieve radios we had not used for years and tuned in to Radio Nepal.

'Dharahara is gone,' they said. Someone said it must be a rumour. But the radio confirmed it. Someone else said, Ghantaghar was gone too, someone dismissed it as rumour. People had died when Dharahara toppled, someone said. Kasthamandap had collapsed, someone else said. Fifty people had been brought into Bir Hospital...and that's how the casualty count began.

In the weeks that followed, all we did was count. We counted the number of people who had died after being trapped under the rubble, people who had died from being hit by falling concrete, people who had died jumping out of windows—we counted like we were sliding beads on an abacus as the numbers shot up every hour by more than scores.

I sat on barren earth with my laptop typing away while the ground continued to tilt and jerk beneath us. Some of the first people I quoted for my stories were people who were right in front of me; others were government sources I called on the phone. First lead, second lead, third lead...the news cycle was never-ending, and in the days to come, would take me to villages that had been entirely flattened.

April 25, 2015 would be etched in memory as the day of horror that stretched into a night of constant trembling, punctuated by women and children crying in fear. Against the backdrop of it all, I continued to talk to people and type endlessly. 'News alert: Update coming up!' I would send a quick note, with details added by my colleagues in Delhi, and further sharpened by editors in Bangkok, editing shifts spilling into the Western time zone as Nepal continued to stagger.

Like most journalists working locally in Nepal, I used to work as a wire correspondent in those days alongside doing a full-time broadcasting job. During the disaster, it turned out that a delicate balance had to be struck between juggling the two acts. I call it balance brought about by work at a time of chaos—perhaps the only way to maintain some form of order.

The TV station where I worked was operating from the street for the sake of safety. The seven-storeyed building, which was our office, had sustained massive cracks. When my brother and I rode on his bike through town to go to our office in Tinkune, where we both worked, we went past crumbling walls, gaping rooms of half-dismantled houses where belongings lay scattered—clothes on beds, plates with unfinished food. Life had been abandoned in stasis.

Everywhere we went, people were sitting in groups in whatever little open space could be traced in the Kathmandu squalor. No one believed it would pass, and for weeks, it did not.

At Tinkune, the OB (out broadcast) van was parked on the street, flanked by a large tent which functioned as a newsroom. When the tent caved in under heavy rainfall, we would drain the water with a stick; when storms hit, we would cling to the ends of the tent. Dry instant noodles became our staple as we updated news by the hour.

With everyone camping, did TV really have an audience? Regardless, some of us wore make-up (some did not), got behind the camera, interviewed and reported. The country had come apart brick by brick, slowly tumbling down into a big pile of infrastructural, emotional, psychological and political debris for everyone to clear up.

'I know this is a hard time, but we're proud of the stellar work you're doing. All journalists live for moments like these,' Thomas, one of my Bangkok editors wrote to me. In Nepal, our news chief was stressed trying to pull together information from across the country, while also making sure no one was risking their lives to tell a story.

Sometimes, we would yell at each other across the tent of a newsroom. Looking back, I am not at all surprised at our loss of tolerance and patience at such a difficult time. Because that is what we do in most newsrooms, remain stoic and get our work done. So, sometimes, you swear under your breath. I think we were actually foul-mouthing nature at that time, even though the words ended up like darts on colleagues.

Our plight was caught on camera by an Indian TV channel that produced a feature on how a Nepali TV station was operating out of a tent. And although journalists should never become news, at times they do. Three weeks after the earthquake had struck, a German radio station asked to do an interview with me on the coverage.

I did the interview by phone, standing outside one of the office buildings. When the interviewer asked me if I had suffered personal loss, I said 'No' at first, and just then, a relative came to mind. She had been in the shower at her home in Balaju when the earthquake struck. Her husband was pulled out alive on the same day, but she had been trapped much deeper under the debris. Finding an excavator to dig up the place had taken two days, and by the time they found her, she was still.

It was only when I answered that my own loss sank in. During three weeks of counting the number of people who had died, I had not realised she was also in those numbers. I had not paused for a moment to think of her. As I responded to the interviewer, I started choking on tears and the interview was wrapped up. I was finally grieving.

From the time the disaster hit, I had spared myself no time to think of the loss experienced within the family. It was deliberate on my part. I had turned myself into a news-churning machine because it seemed like there was no other way.

For weeks, we slept with the main entrance to the living room wide open and every time there was an aftershock, we rushed outdoors. All of us. For me, as it might have been for all journalists, I couldn't afford to think of safety alone. I needed to run outdoors with my computer tucked under my arm every single time. So, the temporary sleeping arrangement included having my laptop by my side, my glasses perched on it, shoes by my feet.
Rattle.
Slip on shoes, pick laptop and glasses and run outside.
Repeat.
It became muscle memory over the weeks.

I was doing much of the news-writing from the porch (sometimes from cafes)—it was where I was calling the Home Ministry and district officials from. It was from where I was being crass, annoying the army officials with queries about the number of bodies they had picked up from the mountains. It was where some of my neighbours, who did not want to be in their own homes, came along with their children to sleep. It was also where my dog sat next to me patiently while I wrote news features.

My brother would take us around on his motorbike when I went out to work. In many localities, we met families sitting in temporary homes, sometimes built on the rubble. In some homes only the children had survived. In some, elders had never had the chance to leave the resting corners of their homes. Deep fissures had suddenly appeared in people's lives with their loved ones and property gone.

Years after those weeks, when I think of April 2015 now, I can only see a feeble shadow of myself, sitting around with a laptop, making calls, walking about town or travelling, talking to people only about loss. I do not see the person living with ageing parents who were traumatised and needed care. I do not see my one-week-old nephew, who was constantly rocked to sleep by the tremors. I do not see our ancestral home in Naradevi with its crumbling walls that I have never visited since. I only see a person who was in a self-induced state of inertia, from the terrain of which I got my job done.

Breaking news, leads, follow-ups and features and not a wink of sleep—I was at the end of my tether by the end of the fourth week. Yet, I did not stop writing because it was the only way to calm my nerves. I had convinced myself that as long as I kept tapping at the keyboard, I did not need to be afraid; in my mind, it helped me remove myself from my immediate situation.

In 2016, I joined BBC Media Action, where I had the opportunity to attend a session on trauma. One Ms Kaur walked us through the impact of working during crisis and helped us talk about our experiences. It occurred to me then that during the 2015 earthquake, no support of that kind had been made available to those of us working in the local media, where resources are always scarce.

Introspection during that session with Ms Kaur helped me understand how I had shut out the reality around me during the earthquake so that I could remain in my bubble and manage to keep working. I had received no psychological counselling, to see me through that trauma. And a memory flashed in my mind. In September 2015, when I had stepped off the plane at Dulles Airport in Washington DC, the ground under me had felt wobbly, filling me with fear that the earthquake had followed me all the way. I had carried my trauma with me across the globe. Over the months.

Inside, I had never stopped trembling.

DOLAKHA

Kanchhi Shrestha, her husband, Bam Bahadur, and seven other family members live in this new house made of iron sheets and scraps of wood from the house they lost.

AARTI
SURAKSHA
IS: 277

Kevin Bubriski: Photographer's journal

Sunday, 7 June 2015, Shaftsbury, Vermont, USA

After the graduation ceremonies at Williams College in Massachusetts earlier in the day, we are hosting dinner in our house in southern Vermont for Nepali friends: graduate Apoorva Lal, his sister and parents, and graduate Avishek Shrestha and his parents. Avi's father, who had arrived in the US from Nepal just two days before, said to me, 'You know, the Nepal you know is no longer; it's gone.'

Saturday, 13 June, Farmers' Market on Kantipath–Bhaktapur

Yesterday, Ian Alsop had taken me on his motorcycle to the village of Halchowk to the west of the Ring Road and Swayambhu. Ian is helping fund the relief effort in this village where he and his family used to live earlier. He also tells me of the Saturday morning Farmer's Market off Kantipath where I am this morning catching up with expat friends I haven't seen for years.

It is a bright hot day with a deep blue sky as I move through Bhaktapur, photographing the dusty work of homeowners picking apart their homes and buildings, brick by brick, sorting and stacking the good bricks from the broken. Large excavating machines are scooping up bucket-loads of debris and pouring it onto waiting dump trucks. Sunlit dust hangs lazily and thick over the streets. Overheated and weakened with a fever, I take a long break in the cool shade of an empty restaurant courtyard at Dattatraya Square. From there, I pick myself up to walk down to Hanumanghat to see the large stone Buddha statue *(p. 108)*, as I had seen it weeks before in news reports, undamaged and unfazed by the devastation and rubble around it.

Monday, 15 June, Patan

It's 12:30 pm under the shade at an umbrella-covered table at the Third World Café overlooking Patan Durbar Square. There are long benches on either side of the Patan Museum's beautiful doorway, guarded by two placid-eyed, all-seeing lions with their beautiful long manes in endless spiral rivulets finely carved into the stone many centuries ago. To the left of the door the bench is tightly filled with older gentlemen in Nepali *topi*[1], watching, talking and listening as the occasional pedestrian goes by. The pigeons are in a spread-out arrangement—motionless on the brick plaza. To the right is the stump of the Char Narayan Temple with its stone *shiva lingam*[2] and square-shaped *yoni* *(p. 127)*. The three layers of its plinth are still there, and the guardian lions are standing upright although off-kilter; gone are the temple walls, struts, tiled roof, and glory. The *shiva lingam* is protected by a square piece of orange fabric with red eaves that move when the wind is strong. Though I've known Patan Durbar Square for decades, I don't specifically remember that temple. In the first evening as I walked through the Durbar Square, I felt that pieces were missing, that there are empty spaces, negative spaces, holes in the sacred configuration.

But the *shiva lingam* is there, is worshipped and receives offerings; it just doesn't have the beautiful home it had had for so many centuries. In my first evening in the Durbar Square, there were many people with phone cameras, all Nepalis, exploring the new post-earthquake square, chatting about what was lost, what was different. There are many Nepalis who must feel the need to bear witness to what has been lost and what has changed. The stone Krishna Temple is standing but surrounded and supported by thin, long wooden struts that look like toothpicks compared to the heavy black stone of the temple *(p. 126)*. I am exactly at the height of the Krishna Temple's third storey where I've sat almost three hours and only now become aware of its total beauty and the intricate complexity of detail of its fine figurative sculptural reliefs carved into the stone—the carvings on the second storey provide a long narrative of the *Mahabharata* and the third storey of the *Ramayana*.

Wednesday, 17 June, Techo, Bungamati

Today, I've lost my overwhelming idea of the romanticism of ruins. I loved reading Christopher Woodward's *In Ruins* and remembering my Peru and Syria experiences through his lens and perspective of respecting and leaving ruins untouched as places in which to vicariously travel back through the imagination to an earlier time.

There was no thought of romanticism when walking into Bungamati. The crown jewel of the village is the ancient home to the Newars' patron deity, Rato Machhendranath. That temple is completely gone. The space of the temple

1 'Topi' literally means 'a hat'. The multi-coloured and -patterned Dhaka Topi and the plain, black Bhadgaunle Topi are headgear that are commonly worn by men in the mid-hills region of Nepal.

2 A votary object symbolising the god Shiva in the shape of a smooth cylindrical mass that is often found resting in the centre of a lipped, disk-shaped object, the *yoni*, an emblem of the goddess Shakti.

square is an endless expanse of rock piles, four to five feet high with a blue police checkpost tent at the back. At the entrance to the temple are some small banged-up painted lions in stone, the bell on its stone lintel frame is still standing, and then a sea of stacked stones *(below)*. Around the temple space and its inventory of piled rocks is a well-worn camp area—a mix of tents, tarps, broken timber, and then across the expanse several tall buildings three to five storeys in different states of having weathered the quake.

Friday, 19 June, Swayambhunath, Gongabu Bus Park

A taxi takes me right from the Bhimsen Temple in Patan Durbar Square through the early morning rain to Swayambhunath. On the way, we buy a nice umbrella for only 300 rupees (less than U$2). It is refreshing and relaxing to be walking up the familiar steps at Swayambhu, up past the brightly painted stone Buddhas, and through the soaking wet greenery in a gentle rain with my new black and silver umbrella overhead.

I pay my 200-rupee fee at the little cement booth, then finish the last 30 or so steps of the climb. The stupa is as ever perfect and in fine shape. Just to its left, the *shikara*-style temple is damaged and as I work my way around the stupa, more and more damage is apparent—shops, residences, and the yellow monastery. The yellow monastery is busy with young men and women filling *doko*[3] with broken bricks and debris *(p. 204)*. Once the baskets are filled to the brim with perhaps one hundred pounds of material, each bearer goes around the stupa clockwise to its western side, and then takes the long walk down to below the taxi and bus parking lots almost a quarter of a mile away and dumps the basket.

On the western side of the stupa are extensive blue tarps over the *chaitya*[4] behind the Ajima (Harati) Temple, believed to protect children from smallpox and other diseases. Peeking under the tarps I see a young boy sitting on a makeshift wooden plank platform wedged between the small *chaitya*. He gives me permission to step into the tent and make photographs of him, solitary in the dim green blue light beneath the expansive blue of the tarps. His parents arrive and are agreeable to sitting on the wooden

3 A *doko* is a conical bamboo basket common in Nepal carried with the help of a tumpline braced against the forehead.
4 The domed stupa-like structure found all over Kathmandu Valley usually constructed in memory of a loved one who has passed away.

Bungamati: Rato Machhendranath Temple

bed frame they have shared as a temporary home every night for weeks now since the quake. They tell me there are 27 families in similar circumstances up at Swayambhu.

Further, I see the silhouettes of almost a dozen Nepali Army soldiers on a rooftop *(pp. 26–27)*. One soldier wields a sledgehammer while others watch. I join them, and we are just at the height of the top of the stupa's dome.

Back at the yellow monastery, I notice a partially standing prayer room through one of the windows and crawl into this open area with no ceiling and no external wall, only rebar and concrete window frames with their multi-coloured decorative details *(pp. 18–19)*. The remains of the windows frame the trees and the urban sprawl below. The inner walls have painted surfaces and there is the wall where statues used to live in glass-covered cases. The floor is littered with broken bits of coloured plaster.

Above me, a few workers with rock drill, jackhammer, pry bar and hammer hack out the monastery's two heavy metal red doors. There is now a moving column of 20 to 30 young workers either shovelling debris or carrying it away. From the sad filth and dust in the stupa neighbourhood, I take a break near Manjushri Hill in the cool breeze beneath the large-limbed trees, rich with green foliage and countless strings of prayer flags. Sitting outdoors at a tiny soda shop, a large monkey sneaks up behind me and makes a serious grab at my bottle of Coca Cola, actually gripping the neck of the bottle tight for a moment. Quite startling.

I get a taxi to the Gongabu Bus Park with the idea of briefly photographing the collapsed big concrete hotels in the area. The young guys removing debris point up an eight-storey building where a handful of workers wave to me from a broken-out window. They motion me up, so with a few of the workers I head up the eight flights of stairs deep in the heart of a quake-destroyed building *(p. 48)*. The heavy thud of a worker's sledgehammer on a concrete wall or floor far above reverberate in the stairwell as we climb up and up. An eerie feeling takes over going up into a structure abandoned yet seeming so solid and so very huge and immovable. On the top floor, a cool breeze blows through the open space where over a dozen workers man a couple of jackhammers, sort bricks, fill baskets with debris, and make the long descent with each *doko* load.

Photography is agreeable to all and I work to get a good portrait of each worker and write a bit of biography as well. This becomes a bit of a distraction from work and soon the contractor says: 'Enough—everyone has to get back to work.'

He, Kumar Rana, is a serious diminutive boss and wants work to get done. He is the owner of the Namaste Home Guest House, one of the high-rises in the neighbourhood. But these days his business is taking down three of these with his crew—sledgehammer blow by sledgehammer blow, *doko* load by *doko* load.

Being up in the air in the open skyscraper becomes somewhat familiar before the long descent reminds me of where I've just been. I walk out into the streets a block from the bus park which is lined with hotels for travellers coming to and going from Kathmandu. The story I hear is that these started as small hotels with little money and then as they started earning they added floor after floor. The upper floors were stronger because there were more resources to put into them. When the earthquake came, the lowest floor could not support the buildings and several collapsed.

I wander into some alleys among the collapsed buildings, feeling the claustrophobia of being in a place with little escape. But there are other people about and one can see where others have walked as well. The rational mind says there is little likelihood of further collapse at this moment, but I also keep track of how to retrace my steps quickly.

I'm drawn to one building that has been quite thoroughly hammered out so that all that remains are the concrete post and beam elements with all floors and ceilings reduced to just the rebar mesh and one can look up to the sky through the floors. The work crew there returns and invites me on board this tilting, listing four-storey shell. Once there and back at work, two of them are two storeys above us hammering away with a jackhammer silhouetted by the bright sky above *(pp. 55–56)*. While I get a few good portraits as well, I don't want to interrupt their flow of work getting names and biographical information.

Saturday, 20 June, Kathmandu

From Kantipath I and a friend walk into old Kathmandu with my first stop at the Chusya Bahal, where the ancient wood *torana*[5] is absolutely as perfect as ever. I haven't seen it since I photographed it for the power places book[6] back in 1989. I remember studying the beautiful rich detail in the printed transparency in book form and realise today that it really is easier to see and read the details in photographic representations than in the sacred object at the site. But I

5 Ornamental gateway leading into Hindu and Buddhist shrines in South Asia.

6 With Keith Dowman, *Power Places of Kathmandu: Hindu and Buddhist Holy Sites in the Sacred Valley of Nepal* (Rochester, Vermont and Thames Hudson, London: Inner Traditions International, 1995)

Dolakha

can touch the scrim of dust on the wood-carved figures of the *torana* and feel the immediate presence of ancient deities on this open street in old Kathmandu. I have had this same sense of being witness to the spiritual and aesthetic resonance of place every moment I've been in Patan Durbar Square. There is a sense of urgency to take note of the beautiful art, not to just walk past, not to take it for granted, but to embrace it in the moment.

I walk through the old city, ducking into narrow passageways and feeling as never before in Kathmandu a deep sense of claustrophobia and almost reluctance. Old-city Kathmandu has always been funky, but now with the fallen structures and the dust and soggy filth, it is off-putting and feels unwell and unhealthy. And then knowing that the end of one's perambulations no longer takes one to the perfect pristine and complete architectural space of the Kathmandu Durbar Square *(p. 6–7)*.

Not far from Asan we watch a crew tug at a steep slope of filthy debris and above this monumental wall of debris are two high-rise narrow buildings of eight or nine storeys. We wander into a large open area off Asan where two sisters and their mother pull bricks from a mountain of debris. They are getting the usual 800 rupees a day and have come from out of town to make money.

Asan is crowded but peaceful at this Saturday midday hour. We head towards Indrachowk and discover the Seto Machhendranath Temple entry closed off. The temple itself is okay but the building with the entrance has cracks and is unsafe as a passageway to the temple.

Sunday, 21 June, Boudha, Father's Day, 6 pm

A dog barks and is echoed by others.
The prayer *Om mani padme hum* floats from the CD shop below.
Crows caw loud, almost as loud as the dogs.
Bells of evening worship are dim in this cacophony.
The Buddha's eyes watch through the thick screen of bamboo scaffolding,
Like a brilliant toothpick sculpture with invisible glue.
The stupa flagless, colourless, waiting, healing,
Ready to rebuild and gather strength.
The sky is heavy with monsoon clouds.

Monday, 22 June, Kathmandu–Chautara

The Polar fan swivels towards the far corner and then to the bed. There are crickets loud enough to hear above the hum and rattle of the fan. We're in Sun Kosi Beach Camp, a bare-bones resort that is now a base camp for European and Irish relief workers.

Most of the afternoon is spent in Chautara, the district centre for Sindhupalchowk district. The ridgetop town of Chautara suffered lots of slippage of buildings that may not have had adequate firm footing on the steep slopes along the ridge. Photographer Chemi Dorje has joined me on this week-long road trip to Sindhupalchowk and Dolakha. Chemi and I first met earlier this year in Dhaka, Bangladesh, at the Chobi Mela International Festival of Photography. He is among the very best of Nepal's many excellent young photographers. Chemi and I walk from town to where a backhoe is pushing debris off the slope *(pp. 166–167)*. Above that is a damaged large shell of a building with exposed interior bright green walls surrounded by greenery, and along the footpath a perfectly untouched gold-painted Buddha statue on a multi-coloured vertical striped pedestal, all unharmed and surrounded by broken bricks and debris *(pp. 168–169)*.

Chautara is a geographically and geologically precarious location, with the lack of zoning evident. So many buildings lost their anchor and drifted from the foundations. As in all the quake-affected areas, it feels hopeless, but, of course, the Nepalis are going forward. Several rock or brick and mud mortar houses are under construction, while the large tilted concrete buildings will linger far into the months ahead before they are picked apart.

Here, at Sun Kosi Beach Camp the heavy, silt-laden, grey-brown churning water of the Sun Kosi river flows 20 feet directly below us sitting in the yard under a very hot sun, blocked at moments by monsoon clouds. The river is silent here, but there is a rustle of rapids a quarter mile away. Monsoon is just beginning, and one can imagine how much this river will swell.

The landscape is lush jungle and moist this time of year. There is barely any traffic on this main highway that links Kathmandu and Tibet/China. Landslides closed the highway after the earthquake, then it re-opened. All the customs and immigration offices and all other buildings in Tatopani at the border were wiped out by the landslides—everything gone. The Chinese opened a narrow passage to

Halchowk (west of Swayambhu)

bring in emergency relief just after the quake and one sees Chinese tents everywhere. But nothing else is crossing the border now due to landslides, and the road is hauntingly and beautifully silent.

Tuesday, 23 June, Sun Kosi Beach Resort–Andheri–Lamosangu–Barabise–Muray

By 9 am we're on the road under a very hot bright sun in our powerful four-wheel drive. We make a stop in the little town of Andheri, where a green building with white-and-pink detail and a dangling serpentine-like balcony railing catches my eye. We end up in a concrete building that is being taken apart. It seems to be part of my work to get out of bright sunlight and into the interiors where the destruction literally surrounds one.

We're in Lamosangu in harsh bright light that is a challenge for any photography, yet the destruction here is incredibly thorough. The highway is narrow with piled debris and broken buildings lining both sides of the street. With the bright light and heat everything feels oppressed and depressed. We meet a handful of young, hip college-age guys with a guitar, cool haircuts and daypacks *(p. 207)*. They are volunteers going out to visit village schools to do art camps and play music. We follow them to Barabise, the large town straddling the Sun Kosi that has suffered quite a bit of damage. Briefly, we visit the resettlement camp above the town and take a few pictures around the Community Anti-Trafficking tent where there are young mothers, babies and toddlers.

From Barabise we head upriver a little way towards Tatopani with the idea that we'll hike up to Gati, a large Tamang village where we'll deliver the books, pencils and pens we've brought with us. But the landslides are impressive and becoming active with the rains that have just started. We wisely decide not to take the risk of going further up the road, and, in fact, there is a bulldozer clearing landslide debris not far beyond Barabise, so further travel upriver is impossible.

South of Lamosangu we have a super lunch at a large place with endless clean, white tables, and white, double-wide benches. While we're sitting afterwards, Shiva, our driver, gets a phone call with news that huge rains north of us have created a flash flood that has swept over the big bridge at Tatopani. When we reach the bridge to cross the river and get on our way to Charikot, there are clusters of locals on their phones, watching the river in anticipation of a flood surge.

Quickly, we climb several thousand feet on a couple of dozen hairpin turns, from the river valley to pine-forested slopes. The air is distinctly cooler. Most of the structures up here are made of corrugated metal sheets and the side of the road is filled with school kids on their way home at 4 pm in their maroon pants and white shirts for the boys, and maroon skirts and red sweaters for the girls. As for the metal sheet structures, it is hard to know whether they are all new or have been part of the landscape for years. Perhaps they replace homes that have been destroyed or people have migrated to the roadside from the steep slopes.

Our brief stop in Thulobagar gives us our first taste of the high-altitude chilly wind, a welcome change from the heat down in the Sun Kosi Valley. There is a rich complex of white-painted stone *chhorten*[7] and prayer flags. A few of the *chhorten* had tumbled in the quake, and one large platform house had also broken. Otherwise, Thulobagar looks okay. I am shown the cement *gompa*[8] which has been under construction for 15 years and its still-rough concrete floor, walls and ceiling appear to have weathered the quake fine. The Tamangs provide a rich visual treasure here with their *chhorten* and flags, and the quick transition allowed by the motor road from the valley heat below makes it even more startling.

We get into thick fog and heavy rain winding through the forest on the black-topped road. A half hour later in hard rain we get to Muray and wisely decide to spend the night there rather than push on in the rain to get to Charikot, which we know is an upended town filled with broken buildings, with no hotels or facilities for folks arriving at dark.

We quickly decide to take rooms on the upper side of the road rather than on the steep side. As we drink tea in the dusk light, the fog envelops the road while kids emerge with a soccer ball and the quiet main road becomes a 'futbol pitch'. The fog and street activity are mysterious and beautiful. Even with drizzle in the air, the feeling is light-hearted. Rain, drizzle and fog continue until dark.

At 8 pm we go across the street to the hotel where our *dal-bhaat*[9] is waiting. Our hotel rooms are dark as the whole town is without electricity. By the light of a battery-operated lamp, we eat delicious rice with two vegetables, three *achars* and chicken with plenty of rich gravy. While we eat, the mother joins her young daughter and toddler daughter on the foam pad and bedding she has unrolled beside us. In

7 A *chhorten* is what a stupa is known as in the northern parts of Nepal where the influence of Tibetan Buddhism is strong.

8 A monastery.

9 The Nepali staple of rice, lentil soup and curry.

the near-darkness of the single dim bulb she falls asleep. It's 9 pm now. We get candles for our rooms and head across the street back to our simple Rs 500/night rooms.

Wednesday, 24 June, Muray–Charikot

We wake up by 7:30 am to steady rain, fog and the news that the road ahead to Charikot is blocked by a landslide and that a 'dozer'[10] is on its way. Shiva has been sipping his tea and watching the empty rain-soaked main road. No kids playing soccer and no vehicles up or down the highway—maybe it's too early, maybe we're locked in between landslides. We finish our tea, chickpeas, potatoes and sweet *malpua* bread and wait for news about the road.

At 9 am, we get the news that the road is open. It's exciting to be on the road in the mist through forests with tall trees and large ferns that cling to the vertical mountain sides. Kids are on their way to school towards town, the road is dotted with new post-earthquake corrugated sheet structures, new homes for the predominantly Tamang and Newar communities who have lost their houses.

We stop at Khare Dunga where about 40 jovial Punjabis have set up camp and are building a very solid large metal-framed, metal-roofed community centre with a cement floor. They are 20 days into the project and plan to finish in 10 more with the building outfitted with rooms and furnishings as well. They offer us tea, biscuits and *laddu* sweets. We're given the tour of the cooking area, supplies, and also catch the head cook taking an outdoor bucket bath, his body covered in soap suds which he washes away with a full bucket of water poured from his head down.

In the neighbourhood are many newly built low wood and tin sheet houses. We chat with a few of the new residents. The story is the same: their houses were destroyed, and they've built these low houses well rooted in the earth to get them through the monsoon. No one we meet is extraordinarily depressed, but instead active and moving forward, building more shelters, going out to harvest potatoes. Of course, there are the squalid tents everywhere and damp places where the dust of a lost house is now wet with the monsoon rain. And many village people are toiling at cleaning up such places so that they can re-inhabit them. It's all dirty work in the ruins of a house. And then there are so many people crammed into makeshift tarp homes.

We get to Charikot in the bright midday hot sun and walk about to make first images of the collapsed and twisted large multi-storey concrete buildings. Due to the blinding bright overhead sunlight, Chemi and I both remark how this was a perfectly unideal time for photography on the street. There is a crew of five or six Oregon State University engineers conducting a survey of the structural engineering of the buildings, and they, too, are taking pictures.

A cluster of young women nursing students see us and open their classroom window to talk to us. They invite us into the classroom where they then cajole us into staying for a short dance performance of theirs. The blue walls of the room echo their blue skirts to perfection. They are so fearless, bold and fun. There isn't a trace of shyness, and their dance is a delight and another ray of hope—women studying medicine and public health among the ruins of Charikot *(pp. 176–177)*.

Dolakha, an old Newar town a few kilometres away, is very sad with its widespread damage. It is a funky old town but now half the buildings have suffered major damage and these grand three- and four-storey brick buildings will certainly crumble in the monsoon rains. Various family groups clean up the ruins, digging out the damp dirt, and rebuilding mud and stone walls. The Dolakha Temple is in fine shape and appears undamaged.

It's late afternoon when we get back up to Charikot. The wind has picked up and the rain has started. We drive the length of town to the Thakali Hotel and Lodge where we'll stay the night. Shiva, our driver, is afraid to stay in the centre of town with all the unstable buildings. The Thakali Hotel and Lodge is owned by an energetic and forceful 57-year-old handsome Chhetri man. He speaks non-stop about his daughters and son in Sydney, Australia, but also talks about the earthquakes. We have reluctantly checked into this seven-storey tower perched on a vertical wall at the edge of the road into town. It's an exercise in acceptance of events larger than oneself. The consolation is that I'm on the second floor and Chemi and Shiva on the third, so we supposedly can make an escape if a quake comes while we are here. Later in the evening, two seismologists—Czech, but working at Oregon State—arrive after a long, harrowing road trip in which their vehicle got caught in a landslide. They called in a backup vehicle from Kathmandu. Asked about the safety of our hotel, they say, 'There are no cracks in the walls. We just have to hope for the best.'

Thursday, 25 June, Charikot–Melamchi

By 8 am, Chemi and I are walking the length of town, revisiting the sites of the large buildings that have slipped, compressed or crumbled. Anisha Shrestha is walking

10 In Nepal, all earth-moving machines are known as 'dozers'. The backhoe loader is the one most commonly used to clear roads, and that is what the dozer refers to here.

with her mother and brother through a major ruin with its pink and white pillars and blue fallen staircase. The family lives right across the street and were unharmed. Their building is still standing, but the brother and sister walk through the ruin twice a day to get to and from school.

A backhoe has just arrived across the street to pull up all the notebooks and documents from the collapsed traffic police station. Four or five well-dressed policemen in freshly pressed blue shirts and white caps with black trim are stacking the wretched wet documents that two young men in shorts and flipflops pull from the wreckage the backhoe has loosened up. There is nothing of value, nothing retrievable, just a soggy sad pile of decomposing paper and refuse.

We pull into Melamchi just at dusk. It looks like a war zone with the entire lower end of the town's old houses all in ruins. The road has also experienced a recent landslide at the other end of town with rocks almost the size of small cars pressed up against the buildings on the far side of the road. Boulders litter the road. A crowd of men are on the porch area of the building with the boulders resting against it. It appears they are taking great risk to be there.

We take an alternate road from town to get to Melamchi River Side Inn. It is just about dark now, and a large bus and a four-wheel drive have brushed up against each other, closing the road. We leave the vehicle and walk by numerous relief tents and kids playing soccer in the dark. It feels raw and tough for sure. The hotel is fine although it is quite hot down by the river. The hotel management has already plastered its walls to cover the cracks on the walls, but they haven't yet hidden the plaster work with fresh paint.

My room is on the second floor and it is very hot. The fan works well, but I don't want to open the windows due to the mosquitoes. There are rooms on the fifth floor with air conditioning, but none of us want to be up that high in a structure with so many cracks already. It is dinner out under a pavilion by the river with its energetic noisy rapids obscured in the darkness. Everyone here, guests and staff, are glued to their phones tapping away, some drinking beer, not conversing. The elderly owner sits across from me on an upholstered couch and has no smart phone; and mine is upstairs in the room charging. He says he has no use for a phone. He's very youthful-looking, 75 years old, in his shorts, flipflops, T-shirt and *topi*. Whether it is the post-earthquake stress or the heat in the valley or both,

Naya Bazaar

Swayambhu

everyone is in slow motion, almost dysfunctional and inanimate.

26 June, Melamchi–Thimpu–Kathmandu
27 June, Kathmandu–Patan–CIWEC Hospital
28–30 June, CIWEC Hospital, Kathmandu

11 July, Shaftsbury, Vermont–North Truro, Cape Cod, Massachusetts

We get an early 7:30 am start and are on the road to the Cape by 10:30 am. Camping for two weeks requires lots of prep and filling the Thule roof rack and Subaru with everything from tent, utensils, food, blankets, beach towels, bottles of wine, gas stove, pots, pans, dinnerware and more. Now, at 6:45 pm, we are settled in with picnic table set up as a kitchen and dining room with our screened in pavilion over it. The tent frame and shell are all we have for tonight, so we can enjoy the stars on this clear night.

It is refreshing and also very odd to have gotten back from a month in Nepal only 72 hours ago and now be settled in the scrub oak and pines of Cape Cod for two weeks, far from the intensity of Kathmandu traffic, the thousands of internally displaced persons in camps under tarps in the monsoon downpour, and the mountain villages destroyed by the earthquakes.

I spent so much of my month in Nepal under the weather with fevers and discomfort. And now I'm trying to retrieve the experience several days later, knowing it will quickly fade further into memory.

30 June, Khokana

By 2:30 pm, I am in a taxi from Photo.circle and Arun Thapa Chowk in Lalitpur and on my way to Khokana with taxi driver Surya Kumar. Surya does not just park and wait but offers to accompany me and carry my backpack. No doubt I had mentioned to him that I had just had the IV removed from my arm last night and had been released just a few hours earlier from four nights in CIWEC Hospital. We walk all through town past mud-encrusted paths and ruins to the lower edge of town where tarp shelters and metal sheet houses are under construction. Everyone is friendly and happy to have a camera in their midst. Two youngsters, two and three years old, are thrilled to sit on the big double bed frame that had just been moved into the family's new spacious metal sheet home *(pp. 146–147)*.

Another couple with two young kids and very modest means are building a house of split bamboo strips which will let a lot more of the outside to the inside, but they do have mosquito nets, they tell me.

1 July, Bungamati

By 7:30 am I am out the door to Bungamati with driver Surya. We start at the bus stop area with a cup of tea then slowly make our way into the heart of town and its neighbourhoods of lost buildings. Everyone overall is accepting of people with cameras it seems. Speaking Nepali breaks the ice. In our second hour in Bungamati, we find a crew of young men and women working diligently and quickly, like an assembly line, filling wheelbarrows with shovelfuls of debris and at almost a run taking the full wheelbarrows to unload at the edge of town. They are finishing up their hard work, washing up at one of the big, black plastic water cylinders before settling down to piles of *dal-bhaat* heaped on leaf plates on the brick-paved streets *(pp. 158–159).*

3 July, Friday, Boudha–Sankhu–Mahankal IDP camp

Chemi appears at 7:30 am for breakfast before our taxi ride out to Sankhu, east of Boudha. It is a hot sunny morning and refreshing to get out to the suburbs, then the green of rice fields.

Sankhu is quite devastated. There are large expanses of rubble with a few simple buildings or remnants still standing. We enter a three-storey concrete building undergoing demolition—a man with a jackhammer and a tall, strong woman sort bricks in the accumulating rubble. A hole has been blasted through the third-storey floor and the woman shoves the bricks shuffleboard-style across the floor and the bricks fall through the hole. The upper storey is pretty well stripped down to just the rebar; the reticulated grid of shadows covers the walls and floors *(p. 87).*

We head back out into the streets with mud beginning to harden up in the bright sun. There is a lot of ugliness and a sense of depression hangs in the air, even with the blinding bright sunshine. Amidst the widespread, uneven and endless field of rubble, there is a single Quonset hut[11] structure without end walls, and a handful of men sit on plastic chairs inside it. They are friendly and give us little attention or care. I try to imagine what it must be like to live in the ruins of what was your familiar and thriving community. What are they talking about as they sit in the open-air hut? How does one begin to clean up what is now acres of rubble?

11 A Quonset hut is a structure made of corrugated galvanised iron sheets with a semi-cylindrical cross-section.

There is a group of around 25 young Nepalis in orange T-shirts with the slogan 'Re-Build Nepal, Habitat for Humanity' *(pp. 92–93).* They are split into a men's group and a women's group, each just picking away at different heaps of rain-soaked rubble. The sun is blazing hot and it is apparent that these two clusters of young people have only enough stamina to barely stay on their feet and drink water. This is not the energised efficient work brigade of Bungamati, but then these young people are having the enriching experience of actually having their shoes and boots in the muck and feeling the heat and challenge of removing mountains of rubble.

It is interesting that Chemi doesn't take pictures of the young student workers, while I find them to be part of the story. I have a democratic lens where everything is visually interesting, whether young or old, ugly or pretty. Chemi is sometimes on the lookout for the older village men with drooping moustaches and *topi* or old village women with plentiful colour and ornamentation, while I find the punk hip youth equally interesting and representative.

Back at Boudha we have our *dal-bhaat* at Kelsang Thakali Kitchen Restaurant with its perfect views of the stupa just outside the windows. Then we are off to the displaced persons camp right up the road from Boudha next to the Hyatt Hotel. We pass the fortress-like Hyatt with its looming solid buildings that appear like oversized bunkers with a bit of Nepali temple strut and pagoda motifs to take the severe edge off the monoliths. The Hyatt gate and greenery are large and spacious, and then abruptly begins the colourful haphazard patchwork of the hundreds of rough tarp homes. Unlike the international clean-up camps at Chautara with Swiss, Norwegian and other international enclaves consisting of crisp white tents, gravelled, roped-off walkways, and security personnel, the Mahankal camp is an organic collection of disparate and also desperate families from all over the quake-affected mountain areas *(pp. 73–74).* There are a few proper solid tents. Most of the shelters are hand-made tarp structures with bamboo ribbing. Today is only the 3rd of July, just a week or two into the monsoon with the bulk of the heavy rains to follow in the next two months.

While Chemi and I are here, it is perfect overcast weather for photos—neither burning hot sun nor heavy rain. Everyone is friendly and overall welcome the attention of the camera. Due to the good weather, life feels somewhat tolerable in the camp right now; people sleep in their tarp shelters, kids play soccer in a small, open space near the large school tents. Well within the camp entrance a few hundred residents of the IDP camp are lined up with identity cards in hand, waiting for food distribution *(pp.*

70–71). There is no security fence of any kind, so anyone can come in from the busy street bordering it. The simple toilets are very few, all located around the perimeter of the camp—maybe two toilets for every fifty families. One must then imagine what it is like to have to use the toilet at night in a downpour, walking back and forth through the narrow muddy lanes, some narrow even for one person; and then the worries of the young women about male predators watching them.

4 July, Saturday, Bhaktapur

At 1 pm, I am in a taxi to Bhaktapur under threatening skies. There is demolition work on the road to the Bhaktapur Durbar Square, so I walk the last couple of hundred yards to the Durbar Square, pay my U$ 15 entry fee, and get out my passport photocopy so that I can extend my ticket for the week. No longer at the Durbar Square was the large vinyl banner grommeted to the palace wall with the scores of names and photographs of the Bhaktapur victims—young, old, men, women *(pp. 102–103)*.

I head south and down to where I had seen so much devastation on my first fever-ridden visit to Bhaktapur two weeks earlier. The weather is flat now, unlike the bright crisp sun and shadows of my earlier visit when the rubble was dusty and the air hot and dry. Now, the town and neighbourhood are moist with the hard monsoon rains from the night before. It is a late Saturday afternoon and overcast, things are quiet. I spot a young man with a face mask, mud-dirtied clothes, and a wheelbarrow of rubble. After he empties the load down the street, I follow him back into an enclosed courtyard where he and three other men shovel rubble into two wheelbarrows. There are two more men on the roof taking down the upper wall. They work in unison with the older one handling the shovel, a rope tied just above the shovel head, and the younger man pulling the rope taut with each heavy shovelful of broken clay and brick dust *(pp. 112–113)*.

This is one of several houses and buildings I go into and remember my son Ryan's words of caution to me about unsafe structures. It is always a balance of caution and itching curiosity. The biggest leap of faith is deciding that the likelihood of a quake or an aftershock at the moment I enter an unstable building is very remote and I'm in the company of men and women who are putting in 10, 12-hour days in these structures. This one is especially dusty even with yesterday's rain. The combination of humidity, the sweat of hard work, and the ambient dust of this old Newari brick home means that the workers have a light grey patina of sweat and rain-moistened dust on their faces. I spend enough time with them to talk and write down who they are, but also being mindful of not distracting them from their work. The image of each feels well done—well seen and respectfully encountered *(pp. 110–111)*.

As I walk towards Dattatraya Square, not more than a minute from Nyatapola is a large glorious old merchant's house coming down. The five workers on the roof wave me up, come down to unlock the door, now half covered by debris on the outer wall of the house. It is a scramble going up, followed by burrowing under a metal sheet to get onto the roof where the five workers have tethered a large wood-carved antique window frame to a rope and are going to lower it down. The owner proudly tours me through the remaining three storeys of his house with its superb, large carved door frame on the second floor. A steady rain has started, and I have just time to get to the tea shop at Nyatapola Square, sitting on the floor on a *sukul* straw mat, and drying off over three cups of fine Nepali milk tea.

5 July, Sunday, Basantapur and Bhaktapur

I get out of the cab at New Road and am in the short queue to buy my admission ticket to Durbar Square when a policeman on duty approaches me. We chat in Nepali and he smiles and says—just walk in, don't worry about payment. My first stop is the tent camp that covers Basantapur, the big open brick square where in normal times souvenir sellers would have their wares placed on fabric spread over simple wood pallets. It is now a ramshackle collection of solid white-framed tents and the ubiquitous rectangular gable-roofed blue or white Chinese Red Cross tents. This is ground zero of the earthquake's impact on the glorious cultural monuments of Kathmandu and weeks earlier photojournalists from all over the world had scoured the tented camp and neighbourhood for images.

I head to the Durbar Square with its 'Tour Route' signs and white lines painted on the pavement to guide everyone around the temple foundations and piles of wood splinters and debris that are partially hidden in places by the ubiquitous corrugated iron sheeting. This morning, the walls of Hanuman Dhoka and the statue of the red-faced and red fabric-draped Hanuman are heavily surrounded by wooden scaffolding but also open for the public to move about much closer. One never knows when a barricade consisting of a simple line of string will be moved or removed and allow more freedom of movement *(pp. 2–3)*. I quickly walk along Hanuman Dhoka's walls to find the best angles of convergence of the diagonal tangle of beams, struts and braces buttressed against the thick, heavy, tall walls *(p. 1)*. While I appreciate the freedom of movement I'm now experiencing, I'm also quite wary of how

Barabise

immensely heavy these very old, dry, dusty walls are and how much risk it would be for tour groups to be close to these walls.

I'm also keeping an eye on my watch knowing that I have to be in Kunda Dixit's Kathmandu University journalism class at 10:30 am in Patan—a half-hour drive away in morning rush-hour traffic. From the wood braces and scaffolding I go back to the tented camp at Basantapur where a number of buttoned-up shirt and necktie-wearing boys and girls are gathered in one tent—a chance to get 'in tent' shots of school kids.

Now, I rush to a taxi, tell Kunda by phone that I'm on my way, and patiently let the driver take his course. I arrived at 10:27 and join Kunda in a loud echo-filled, glassed-in classroom with my website and Instagram feed up on the screen. The dozen-plus journalism students and Mickey, the videographer from Hawaii, all arrive by 10:43 am. I speak in Nepali for the first five or ten minutes then in English and then back in Nepali. As I show old vintage images from Kathmandu and other parts of Nepal, I talk about the power of photographs to transport us across time and space. I find myself describing my Nepal work, especially my current visit in the post-earthquake period, as how a doctor makes the rounds of the patients to see how they are doing and check up on them, using the camera and photos and stories as a way to stay in touch and put others in touch.

It is almost noon. I am hungry and head to Dhokaima Café. All the umbrella tables are taken by hip, young, affluent, beautiful Nepali couples except for one large table around the courtyard corner. As soon as I sit down and order my fried momos and fresh lemon soda, the rain starts. The couple closest to me start to shift their seats out of the rain and also their laptops to keep them out of the rain. Then the rain falls harder and they along with all the other outdoor customers make a dash for the interior spaces of the restaurant.

I have good cover over my table from the large umbrella and the eaves of the roof as well, so my backpack is safe along with my journal, iPhone and attached power pack. A lone, cream-coloured, short-haired mutt takes refuge in the doorway alcove behind my chair with my backpack. He curls tightly into a ball against the cold wind and driving rain. While I am dry, the rain is building into deep puddles where I am seated, and I find a way to keep my sneakers dry by anchoring my feet on the wrought iron framework

Kathmandu Durbar Square

of the table. The momos are ready and the waiter braves the deluge with a run to my table under cover of an umbrella. I order a milk coffee and a slice of cheesecake. There is no way to not get soaked in this downpour, so my strategy is to sit it out—the lone customer in the rain, sipping milk coffee and having thin slices of cheesecake and writing my journal to keep me busy until the coast is clear.

From Patan Dhoka, at 1:30 pm I am itching to be on my way to Bhaktapur even though the rain is still steady. The entire drive to Bhaktapur is in the rain and when we arrive at the demolition site on the main road to the Bhaktapur Durbar Square, I leave the taxi and give word to the driver I'll be back in two or three hours. Clay dust and dirt has become a slick layer of slippery mud and almost everyone in flipflops was sliding around over the wet muddy piles of debris.

There is an old man in oversized, ill-fitting cloth shoes held together by their laces. He has a walking stick in his right hand and an old hooded poncho draped over him. His body shakes with age and the uncertainty of finding firm footing. It takes him time to decide where each foot should go with each step in the slippery debris and then even in the wide brick landscape of the rain-soaked empty Durbar Square. Who is this older gentleman? How old is he? What did he experience on 25 April?

I head down behind the Bhairab and Nyatapola temples where I have been on earlier visits. This is all about mud and finding footing carefully enough not to slip onto the wet world of brick fragments, wet clay mud, and perhaps rusted metal or glass among the thick mix. I am in a world of soaking wet T-shirt under my waterproof windbreaker and umbrella, my backpack and the blue and white Chinese towel around my neck soaked as well. I keep my camera dry enough to be functional and keep my eyes open for anything. There is a vegetable-and-snacks cart standing against a bright orange tarp protecting a broken temple. There is something very melancholy and beautiful here. Patiently, I wait for someone in bright clothes or a bright umbrella as the counterpoint to the curious shape of the cart against the theatrics of the orange tarp.

Heading downhill, I follow the path to the tented camp at the end of the long street of devastated houses. This is a grimness that transcends yesterday's grim scenes. I step into a deep puddle, so my shoes have gone from very wet to dripping wet. The tents are orderly on either side of the raised walkway leading to a temple. I think of everyone in the tents and hope they are dry and somewhat comfortable. To my left is a road, a very muddy route that will lead to Dattatraya Square. A middle-aged woman in a green salwar-kameez is watching me from her doorway. She is talkative and tells me her house was spared by the earthquake, and then her neighbour is in her own doorway, and her house was spared as well. Later, when I look at the portrait of the first woman on my iPhone I'm touched by the deep sense of sadness or resignation in her eyes and her tired face, and yet remember her animated humour and friendliness as well. How much can anyone trust photographs, or do photographs document the reality our emotions and eyes don't always catch?

The rain has diminished to a drizzle. People are beginning to emerge and make their way up the streets that are covered in mud, and my wet socks and sneakers feel heavy now. I make a few stops along the main street of Bhaktapur on my way to Dattatraya Square. There is a gaping hole extending through the third- and fourth-storey interior wall of a large house *(p. 115)*. It is shaped almost cathedral-like with bottom corners squared and the top half rounded like a Romanesque church archway. An elderly woman sits silently watching me from the front stoop of her house nearby and is joined by a younger woman, perhaps her daughter, who also watches me—rain-soaked, umbrella, pants legs mud-splattered, camera and lens around my neck. I say a few words in Nepali about the rain and the mud. They remain silent. I decide their silence won't forbid me from moving closer and onto the rubble below this extraordinary cavity in the wall to get the photograph, but I feel their presence very deeply and am reminded that I am in their personal space, their neighbourhood, and that numerous photographers, tourists or journalists may have taken the same exact steps to make the photograph.

* * *

Six weeks after the 25 April earthquake I arrived in Kathmandu to make photographs that build on my documentary work of the previous four decades. In fact, my arrival in June 2015 is just a month or two shy of the 40-year mark since I arrived in Nepal for the first time. Questions I ask myself: Does anyone need more images of the unfolding tragedy in Nepal? What can one photograph show that has not been seen or said already? And then the ethical questions of who am I to walk through someone else's neighbourhood where death and loss and tragedy are so fresh; what right do I have to walk on the ruins of someone's home; and did their mother or child die in the rubble under my feet?

Yesterday, at the farmers' market at Le Sherpa Restaurant, in Lazimpat, Brent Wiley, who has been doing independent aid work in Nepal for many years, said many nice things about my photographs and I let him know that it was his recent purchase of some of my Nepal photographs that paid for my ticket to Nepal. He then asked about whether I was doing 'disaster porn' photography in Bhaktapur and elsewhere. This opened up for me a new perspective and questions about my current work—what is publicly permissible, am I practising self-censorship of what I shoot, or self-censorship of what I edit and show? Is the two-storey gaping hole in the wall of this building next to the silent, sullen mother and daughter on the stoop of their house also disaster porn? Is this the scene of the death of a loved one of theirs?

A photographer can be like a scavenger looking for treasure. Admittedly, this is a feeling that frequently haunts me doing this kind of photographic coverage. Admittedly, it was a feeling I dealt with every time I photographed the throngs of grief-stricken spectators at the ruins of the World Trade Centre site from September through December 2001.[12]

* * *

Through the muddy main road, I approach Dattatraya Square. While I wonder how the road could not be shovelled off and cleaned of its layers of mud and dirt down to the clean red brick road, I have to consider how much rubble removal is constantly going on and the general malaise of the community here with such devastation that a mud-free road may seem trivial.

At Dattatraya, shops are open, and the square feels less filled with debris than two weeks ago. The restaurant at the low end of the square is no doubt open, where I had a lemon soda and momos on my last visit when I was in the midst of a debilitating fever. Remembering how weak and handicapped I was with fever seems to be a main theme of this visit to Nepal.

At the upper end of Dattatraya, there is an open neighbourhood tea shop. I have just enough time for a quick glass of sweet milk tea before walking the length of town back to my taxi. Instead, I follow a couple of teenage girls down the narrow alley where the famous wood-carved 'peacock' window sits in the second storey wall of a classical brick-and-wood Newar building. At the end of the block is a broken electrical pole with a tangle of wires and the usual pile of rain-soaked rubble. Just beyond, where three months ago was an extensive neighbourhood of three- and four-storey houses, is now a huge expanse of several acres of complete devastation. The space is wide

12 Published as *Pilgrimage: Looking at Ground Zero* (New York: powerHouse Books, 2002).

open with the skeletal remains of some buildings. The exposed interior rooms painted purple, pale green, blue and orange catch the eye in this grim world of the late-afternoon darkening sky and rain. Scores of houses may have been lost here. While in Sankhu on Friday we were in the blistering hot, bright sunny world of devastation, it is maybe more gut-wrenching here now in Bhaktapur in the wet, dim light. The magnitude of each fills one with a sense of both hopelessness and powerlessness. In both of those similar yet different worlds, no one is picking through the rubble for scraps of wood, belongings or salvageable bricks. Maybe they are, but just maybe the devastation was so thorough that it feels meaningless to be there.

While I clamber about on the slippery heaps of ruined homes and lost lives, I feel like an opportunist taking from others' losses. But I also have little time and work quickly, not limping along, not wanting to be seen for long in this sad place. One young girl walks by, the pink and yellow highlights of her clothes the only dash of colour in this bleak dark landscape of loss. And, then, silence engulfs the space again.

6 July, Monday, Naya Bazaar–Kanchhi Didi's memorial service–Gongabu area

We taxi through the endless high-rise buildings of Naya Bazaar, passing a few notable collapsed ones, and arrived at the place where the *kriyaputri*[13] spend 13 days performing the funeral rites. I sit silently with Bharat for a while. Just beyond our feet is the wide spread of blankets with numerous stitched leaf plates and bowls with varieties of fruits, sweets, *sel roti*,[14] flower petals, red *tika* paste, and a young Brahmin sitting amidst it all, sorting and consolidating all the different plates and their contents. At the front of this array sit Kanchhi Didi's three children, Devendra and Rupendra, the two sons in white robes with heads and eyebrows shaved with tiny white *tupi* stems of hair left at the centre back of the head. Sugunda is in a regular skirt and blouse.

Devendra is back in Kathmandu from his permanent home in Bangkok where he is married to a Thai diplomat. Rupendra is now in his early 40s and he and his wife are full-time physicians at a Kathmandu hospital. I always remember Rupendra from the 1984 photos when he was a little kid dancing on the clay floor of the thatch-roofed brick house of their Newar landlord with its surrounding landscape of rice fields. I remember Devendra best from a 1976 Bhai Tika photo of him, with his huge rice and yoghurt tika, laughing with his father on the railed-in platform porch of their Kharanitar house in Nuwakot district. As I look at the two brothers now, I swear that Devendra is Dil Das Thapa. Both sons look very much like their father, but Devendra more so, while Rupendra still has the young child-like look and demeanour of 31 years ago with an added slight paunch and the beginnings of a secondary chin.

The house in which Bharat's family rented the ground floor was a 100-metre walk through rice fields from the road to Balaju. The jovial Newar family and their beautiful daughters lived upstairs. Bharat's family bought a piece of land from the Newar family and started what was to become a solid 'good up to eight on the Richter scale' reinforced concrete building, and now their house is just one in a sea of hundreds, of which Naya Bazaar is one sea among the ocean of huge concrete buildings that has become the Kathmandu Valley.

Sugunda must be 40 by now. She was always and still is beautiful and has a full-time job at a bank. I would like to speak with her and her brothers, but this is the 13th day since Kanchhi Didi's death and not really a day for catching up with their Uncle Bharat's friend.

Just behind her two sons and daughter is a wood-framed colour portrait of Kanchhi Didi (Krishna Thapa), propped up on a wooden chair with incense and offerings in front of it *(p. 31)*. When I step close to photograph it, I am taken back to her voice, her attractive large matriarchal stature, like the lead in an epic opera. I feel the loss of her as the guiding force of the extended family who survive her: her husband, three children, older sister, Shiva Maya, and three brothers—Bharat, Jit Bahadur and Sashi. I remember her from 39 years ago when she was the beautiful young wife and mother of young sons and daughter, and Dil Das, a handsome young Nepali government worker in the customs office at the China border. Now, I regret I didn't photograph everyone on that 13th day of Kanchhi Didi's death, to photograph who everyone has become since those memorable moments 30 or 40 years ago.

Bharat talks about all the conflict in Kharanitar. While throughout the districts of Gorkha, Dhading, Sindhupalchowk, Dolakha, Rasuwa and Nuwakot, many villages, communities and districts quickly and equitably figured out the household losses and damages from the earthquakes, I am told about some irregularities in distributing the support. I remember my dinner with Austin Lord and his wife, Sneha, a couple of weeks earlier and his specifically remarking how communities in lower Nuwakot were changing the numbers of houses and lives lost, inflating

13 Nearest relatives of the dead.

14 Deep-fried, sweet rice flour bread with a hole in the middle, somewhat like a doughnut.

Charikot

them in order to boost the government compensation amount.

Mid-afternoon, Bharat, Bimala, Alok and I take a taxi back to their house with the idea that Alok and I would go off to photograph 'building breakers' taking down concrete houses in the area. We quickly get to a striking structure with big holes knocked through the roofs and shreds of concrete still clinging to the rebar webbing that had been pummelled and beaten by jackhammers and sledgehammers. And, just as we get out of the cab, the skies open up. Alok and I are indefinitely stuck in this half-demolished shell, wet from the downpour and sweaty due to the humidity.

The half dozen men and two women labourers are amused by our visit and eventually bring us a couple of ice-cold Sprites. We are there almost an hour until the rain lets up and we walk in the hard drizzle to the street past the Gongabu Bus Park with its buildings active with demolition crews. Due to the rain, work is halted except for a three-storey shell of concrete beams and tangled rebar that a steam shovel operator is banging at, trying to get a solid bite with the large heavy hydraulic bucket arm of his bellowing machine. Everyone on the street is standing out of the rain under the eaves and roofs of open shops watching this spectacle. The building's skeletal shell appears to be winning so far.

Absolutely soaked, we stop at a modern shiny café with its Formica tables behind big plate glass windows for coffee and cake, then out into the downpour in the day's diminishing light and diminishing prospects of finding a taxi. We walk and talk for maybe a mile, eventually finding one. Alok gets off at his house and I am relieved to be out of the rain and on my way to change clothes and dry off in my room in Boudha. I put out the umbrella, rain jacket, wet shoes and socks and T-shirt in the hallway and change into a long-sleeved dry shirt and flipflops, and with a torch in hand make my way to the quiet Boudhanath stupa, watching for deep puddles.

It is remarkable how serenely quiet the stupa is on this night of HH Dalai Lama's 80th birthday, but that is the power of the hours of driving rain. I am the last customer of the night at Kelsang Thakali Kitchen for my last *dal-bhaat* with chicken curry, a variety of *tarkari*, and a small cup of chilled buttermilk. As I tell them how good their food is, I have to tell them about the Thakali Restaurant and Lodge in Charikot run by a Chhetri. They get a good laugh.

Chautara

Tuesday, 7 July, Kathmandu–Doha–USA

By 11:15, Alok and I are in a taxi in the rain on the way to a neighbourhood of earthquake-damaged houses and soon find a four-storey building with white balconies and pink outer walls. We find a way over a 45-degree incline of rubble to the second storey and climb out of the rain and into the ruin where Alok speaks to a worker and asks if we can come in and if there are other workers. Yes, we can come in and, yes, there were eleven other workers. We climb up the rubble-covered stairway to the third floor where all the other workers are taking a break. Two pairs of men are lying on their sleeping mats and blankets; one of the four is still sleeping, the other three lay out with earbuds in place, listening to music from their mobile phones while in a room across the stairway is another sleeping space with half a dozen of them finishing up an animated card game. As their attention goes from the completed card game and the call to get back to work, Alok is saying, 'Kevin here is visiting from America and he is a well-known photographer making photos of Nepal for 40 years.'

I understand every word of his and he, like his father, Bharat, proves to be a persuasive spokesman. Meanwhile, I've been making photos of young men stretched out listening on their earbuds, playing cards, and hanging out among the half-demolished interiors *(pp. 36–37)*. It was time to get to work and I follow them up to what is now the roof, but a few weeks ago would have been the floor of the fourth floor. The walls are pale blue, and the workers are slowly getting into their hammer-swinging work mode. While a few are working on the remains of a vertical wall, the young fellow, very handsome, who 10 minutes ago had been under his blanket listening to his earbuds now with great ease and poise climbs on top of a remaining pillar with the bent-over rebar that looks like shrivelled plant life *(pp. 40–41)*.

Alok has taken notes on the workers, describing their clothes and writing down their names and villages. I am reminded of Lewis Hine's heroic 1930s photographs of the proud workers building the Empire State Building. I photograph around six or seven workers and then know we have to start heading out, though I feel this work on the 'building breakers' is not as complete as it can be. I think of Edward Burtynsky's photographs of the 'ship breakers' of Chittagong, Bangladesh. Yet, instead of making the buildings the subjects of the photographs, my photographs use the buildings as the context for the environmental portraiture of the workers.

Skidding our way down the concrete fragment and rubble-covered stairways, we leave the tilting half-demolished building in a monsoon drizzle. I had said earlier to Alok that there is enough photography for me to do in just this one location. I also know that leaving this building is the beginning of the end of my visit to Nepal. This is the far point in my journey. From this point it is the simple narrative of the journey home.

We get back to the taxi; the driver is a family friend of Bharat and Alok's, who has him waiting for us over an hour and a half near the wrecked building. Alok and I get out in the light rain at the CIWEC clinic parking lot for my last follow-up. Alok looks me in the eye respectfully and holds my hand as he eloquently talks about how the two most important people in his father's life are Kanchhi Didi and me, and how he has always looked to the two of us for guidance. Alok goes on to say that now that Kanchhi Didi is gone my role is even more important. I think of how very well-spoken and sincere Alok is, and I feel proud of him and for Bharat for having such a fine young guy as his son.

In the taxi, Alok had confided to me, 'You know, yesterday, when we went into that building I was really scared, but now I feel much more normal about it all.' He got to see me in my work mode and actually spend time with me as assistant, note-taker, facilitator. What it means spiritually, socially and economically to be so important in Bharat's life has me thinking where we go from here. We stand there for several minutes in the rain, Alok speaking to me very touchingly and making me think momentarily that we don't always choose all our friends and there is a beauty in that.

I step into the CIWEC door to have the receptionist tell me that my appointment with Dr Ravi was for 1 pm not 1:30, and that I would have to wait a while to see him. I am now actually glad that I had taken the extra half hour with the building breakers. As I'd said to Alok, in the one building we were in, that we were fine to stay where we were; there was more than enough to do in this one building with its cast of 12 individuals—each a life, all working hard through their physical labour to feed themselves, to save money for their families, to be working, living, experiencing what Kathmandu is. And also living, sleeping and cooking food in the unstable, crooked building they are dismantling with their own hands. This is a story to which I just can't give enough of my attention.

The contributors

Prawin Adhikari writes fiction and screenplays, and also translates. He is Assistant Editor at *La.Lit*, the literary magazine.

Kevin Bubriski first came to Nepal in 1975 as an American Peace Corps volunteer, working through 1979 on remote village community water supply projects in Taplejung, Nuwakot, Mugu and Humla. He returned over the following three decades to photograph Nepal as a recipient of Guggenheim, National Endowment for the Arts, Fulbright and Robert Gardner Peabody Museum fellowships. His fine art photography is in the permanent collections of the Museum of Modern Art and Metropolitan Museum of Art in New York, and Bibliothèque Nationale, Paris. His books on Nepal include *Portrait of Nepal* (1993), *Power Places of Kathmandu* (1995), *Maobadi* (2011), *Nepal*: 1975-2011 (2014), *Kailash Yatra: A Long Walk to Kailash from Humla* (2018), and *Mustang in Black and White* (2018).

Anil Chitrakar is trained as an engineer and energy planner and has been passionately involved in the recovery of Nepal's globally unique natural, cultural and built heritage. He can be contacted at anilchitrakar.com.

Muna Gurung is a writer, translator and educator based in Kathmandu. Besides teaching writing at the Open Institute for Social Sciences, she is the founder of KathaSatha (https://www.kathasatha.org.np) and writes a monthly column for *Nepali Times* called 'Lightroom Conversation'. When not writing or teaching, Muna runs ĀMĀKO, an inter-generational pickle company.

Tsering D. Gurung is a writer and communications professional based in Kathmandu. Her interests lie in covering issues related to human rights, social justice and the media. She has previously worked at *The Kathmandu Post*, *Nepali Times* and Amnesty International Nepal.

Niranjan Kunwar is invested in literature, contemporary art, queer issues and pedagogy. He is the author of a memoir, *Between Queens and the Cities*, and his English translation of Amar Neupane's *Seto Dharti* is forthcoming. His interests lie in actively seeking to counter dominant discourses and practices.

Smita Magar is a journalist and writer interested in rural reporting with a focus especially on gender, education, health, migration and human rights in post-conflict societies. She was a Chevening fellow and holds a Master's degree in Conflict Prevention and Peacebuilding from Durham University, England. She is a founding-member of Jhumlawang Village Foundation, Nepal (JVF-Nepal), Rukum, a community-based non-governmental organisation working for sustainable development of the Jhumlawang village.

Pramod Mishra is from a village in Morang in Nepal and now lives near Chicago, teaching English at Lewis University, Illinois. His articles have come out in various journals, including *The New Centennial Review*, *Ariel* and *Studies in Nepali History and Society*. He has written a regular column for *The Kathmandu Post* since 2009.

Pranaya SJB Rana is a writer, journalist, and editor based in Kathmandu. He is currently editor of The Record (https://www.recordnepal.com), an independent digital news publication, and the author of *City of Dreams*, a collection of short stories.

Avishek Shrestha grew up by the BudhiGandaki river before moving to Kathmandu for school. He visited family every year and spent time by the river at every opportunity.

Deepak Thapa is the Director of Social Science Baha, Kathmandu, and Editor at Himal Books. He has written extensively on Nepal's contemporary history, and has a number of publications to his name, both as author and editor.

Manjushree Thapa is a Nepali writer based in Canada. Starting with *Mustang Bhot in Fragments* (1992), she has written a number of fiction and non-fiction books. Notable among these are *The Tutor of History* (2001), *Kathmandu: An Elegy for Democracy* (2005), *Siklis: The Life and Times of Chandra Gurung* (2009), and *All Of Us in Our Own Lives* (2016). She is also a translator of Nepali-language works, the most recent being Indra Bahadur Rai's *There's a Carnival Today* (2017).

Rabi Thapa is a writer and editor. He is the author of *Thamel: Dark Star of Kathmandu* and the Editor of *La.Lit*, the literary magazine from Nepal.

Kesang Tseten is a documentary filmmaker and writer. His films, screened regularly at international film festivals, include *Who will be a Gurkha*, *Macchendranath*, *We Corner People*, *Trembling Mountain*, and a trilogy on Nepali migrant workers. His writings include articles, short stories, and screenplays for the feature films, *Mukundo* and *Karma*.

Pratibha/Prateebha Tuladhar is a journalist, teacher and communication worker based in Kathmandu. She has worked as a development communicator in Kathmandu, Yangon and Chiang Mai, and has taught writing to photo-journalists at the Asian Center for Journalism at the Ateneo de Manila University, Philippines, for more than a decade. She also teaches journalism at the Communications University of Zhejiang, China, for the University of Bolton. Currently a contributing writer for *Nepali Times*, she has written for the German Press Agency (dpa) and *The Kathmandu Post* and produced short videos for BBC Media Action, National Geographic and *The China Daily*. She writes poetry when she finds inspiration.